Practical Guide to Hereditary Breast and Ovarian Cancer

Daisuke Aoki • Seigo Nakamura • Yoshio Miki
Editors

Practical Guide to Hereditary Breast and Ovarian Cancer

Annual Meeting of the Japanese Organization of Hereditary Breast and Ovarian Cancer 2021

Editors
Daisuke Aoki
Akasaka Sanno Medical Center
International University of Health and Welfare Graduate School
Minato-ku, Tokyo, Japan

Yoshio Miki
Research and Development Center for Precision Medicine
University of Tsukuba
Tsukuba, Ibaraki, Japan

Seigo Nakamura
Department of Breast Surgical Oncology
Showa University, School of Medicine
Shinagawa-ku, Tokyo, Japan

ISBN 978-981-99-5233-5 ISBN 978-981-99-5231-1 (eBook)
https://doi.org/10.1007/978-981-99-5231-1

This Springer imprint is published by the registered company Springer Nature Singapore Pte Ltd.
The registered company address is: 152 Beach Road, #21-01/04 Gateway East, Singapore 189721, Singapore

Contents

Part I
Breast

Establishment of a Medical System with HBOC in Mind

Rika Narui, Kanae Taruno, and Seigo Nakamura

Abstract It has been reported that 5–10% of breast cancers are hereditary, and about half of all hereditary breast cancer are hereditary breast and ovarian cancer (HBOC), which is caused by *BRCA1* (breast cancer susceptibility gene 1) and *BRCA2* (breast cancer susceptibility gene 2) mutations. When HBOC is identified in breast cancer patients, the prevention and treatment options include prophylactic surgery of the breast and ovary, screening for contralateral breast cancer with annual breast magnetic resonance imaging (MRI) surveillance, use of poly (ADP-ribose) polymerase (PARP) inhibitors, and genetic testing of relatives. Therefore, HBOC should be considered when treating breast cancer patients, and genetic testing should be suggested for patients who need *BRCA1/2* gene testing. Once *BRCA* mutation is identified, it is necessary to support patients' decision-making by providing information from multiple disciplines. The establishment of a surveillance system for *BRCA1/2* mutation carriers who do not develop breast cancer will be an issue for future consideration.

Keywords *BRCA1/2* · Mastectomy · PARP inhibitor · Chemoprevention · HBOC · Genetic test

1 Introduction

It has been reported that 5–10% of breast cancers are hereditary [1], and about half of all hereditary breast cancers are hereditary breast and ovarian cancer (HBOC), which are caused by *BRCA1* and *BRCA2* mutations [2]. *BRCA1* and *BRCA2* are tumor suppressor genes that contribute to DNA repair and transcriptional regulation

R. Narui (✉) · K. Taruno · S. Nakamura
Division of Breast Surgical Oncology, Department of Surgery, Showa University, Tokyo, Japan
e-mail: tsukui@med.showa-u.ac.jp; ktaruno@med.showa-u.ac.jp; Seigonak@med.showa-u.ac.jp

D. Aoki et al. (eds.), *Practical Guide to Hereditary Breast and Ovarian Cancer*, https://doi.org/10.1007/978-981-99-5231-1_1

in response to DNA damage [3]. By the age of 70 years, the cumulative risks of breast cancer in *BRCA1* and *BRCA2* mutation carriers are 57% and 49%, while that of ovarian cancer are 39% and 17%, respectively [4]. The risk of developing contralateral breast cancer within 20 years of initial breast cancer diagnosis in patients with *BRCA1/2* mutations is reported to be 40% for *BRCA1* and 26% for *BRCA2* [5]. In the past, the focus has primarily been on breast and ovarian cancers, which have higher penetrance; however, recently, prostate cancer in men and pancreatic cancer in both sexes have also been reported to have higher penetrance than that in the general population [6–8]. The cumulative risks of prostate cancer in *BRCA1* and *BRCA2* mutation carriers are 29% and 60%, respectively [6]. When HBOC is identified in breast cancer patients, the prevention and treatment options include prophylactic surgery, screening for contralateral breast cancer with annual breast magnetic resonance imaging (MRI) surveillance, use of poly (ADP-ribose) polymerase (PARP) inhibitors, and genetic testing of relatives. Therefore, HBOC should be considered when treating breast cancer patients, and genetic testing should be suggested for patients who need *BRCA1/2* gene testing.

1.1 History of BRCA1/2 *Clinical Practice in Japan*

In 1994, *BRCA1* was identified by Miki et al., and in 1995, *BRCA2* was identified by Wooster et al. [9, 10]. Since then, various studies of *BRCA1/2* genes have been conducted worldwide. In 2010, the project "Management for patients with HBOC and unaffected *BRCA* mutation carriers in Japan" was initiated by one of the research groups of the Japanese Breast Cancer Association. In 2012, the Japanese HBOC Consortium was established as a coproject involving breast oncologists, gynecologists, and geneticists. In 2016, the Japanese Organization for HBOC (JOHBOC) contributed to the improvement of preventive medicine by developing and expanding the medical treatment system for HBOC for suspected patients and their families. JOHBOC is associated with various academic societies, such as the Japan Society of Human Genetics, Japan Society of Obstetrics and Gynecology, and the Japanese Breast Cancer Society. The JOHBOC has contributed to the accreditation of HBOC treatment facilities, education and training on HBOC, registration of patients with HBOC, and surveys and research on HBOC. In 2017, the Guidebook for Diagnosis and Treatment of HBOC syndrome was published by JOHBOC. The guidebook recommends the *BRCA1/2* genetic test for patients who met one of the following criteria: an individual with a known *BRCA1/2* mutation in the family, breast cancer diagnosed at age 45 years or younger, breast cancer diagnosed at age 60 years or younger with triple negative breast cancer, male sex diagnosed with breast cancer, two or more primary breast cancers on both sides or one side, blood relatives (within third-degree relatives) with breast cancer or ovarian cancer, an individual with ovarian, fallopian tube, or peritoneal cancer, companion diagnostics for using PARP inhibitor, and *BRCA1/2* mutation suspected by tumor profiling. In 2018, *BRCA1/2* genetic test was used as a companion diagnostic for using PARP inhibitor for patients with inoperable or recurrent breast cancer and human epidermal growth factor receptor 2 (HER2)-negative disease who had received prior chemotherapy. The test was covered by health insurance. In 2020, *BRCA1/2* genetic test for the diagnosis of HBOC was covered by health insurance for those who met one of the aforementioned criteria. Genetic counseling and surveillance of contralateral breast cancer with annual breast MRI were also covered by health insurance for breast cancer patients with *BRCA* mutation. Since *BRCA1/2* genetic testing has been covered by health insurance, the number of *BRCA1/2* genetic tests and contralateral risk-reducing mastectomy (CRRM) is on the increase. The number of *BRCA1/2* genetic tests increased from 54 in 2019 to 126 in 2020 in the breast center of Showa University Hospital; however, the data were obtained from a single facility. Approximately 28% of unilateral breast cancer patients who had *BRCA1/2* mutations between April 2020 and March 2021 in the breast center underwent CRRM. In 2021, the HBOC clinical practice guidelines were published by the JOHBOC. These guidelines are expected to increase testing and treatment related to *BRCA* in the future.

1.2 **BRCA1/2** *Genetic Test*

BRCA1/2 genetic testing is usually performed on patients with a positivity rate of 10% or more [11, 12]. There are some international guidelines and recommendations for genetic screening of *BRCA* [13]. In the United States, the National Comprehensive Cancer Network (NCCN) guidelines provide testing criteria for *BRCA* genetic testing [14]. Yadav et al. in the United States reported that approximately 47.9% of women with a diagnosis of invasive breast cancer (84.0%) or ductal carcinoma in situ (16.0%) met the NCCN *BRCA* testing criteria [15]. Cropper et al. in the United States reported that the positive rate of the patients who fulfilled more than one NCCN testing criteria was approximately 10% [9]. In Europe, the European Society for Medical Oncology (ESMO) guidelines provide *BRCA1/2* genetic testing criteria [16].

In Japan, HBOC clinical practice guidelines were published in 2021. According to the guidebook, the criteria are considered to be odds ratio > 2 or *BRCA1/2* mutation detection factors of 10% or more from the data of Japanese breast cancer women without selection bias in the NCCN guidelines and the Japan HBOC Consortium's simple check [17]. There is no report yet on the exact positive rate when this is applied to unbiased patients; therefore, further studies are necessary. Guo et al. reported that the selection criteria for *BRCA1/2* testing and genetic counseling have gradually loosened over time [18]. There are some reports that *BRCA1/2* testing is moving toward a wider range of subjects in the future [13, 19].

The number of *BRCA1/2* genetic tests is increasing in clinical settings, and there are some future issues. First, multidisciplinary support is needed for decisions within a short period from the diagnosis of breast cancer to surgery (Fig. 1). A patient with a *BRCA1/2* mutation must decide whether to undergo prophylactic

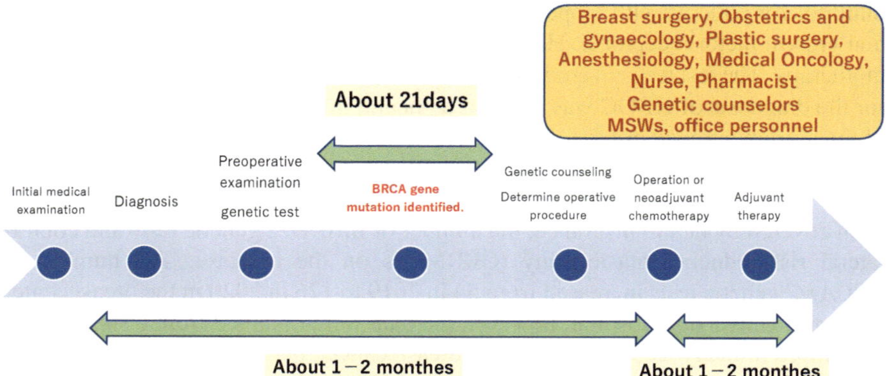

Fig. 1 The flow of breast cancer treatment when *BRCA* testing is performed. Multidisciplinary support is needed for decisions within a short period from the diagnosis of breast cancer to surgery

surgery within a short period prior to surgery. Therefore, it is necessary to support patient decision-making by providing information from multiple disciplines, including doctors, genetic counselors, nurses, plastic surgeons, obstetricians, gynecologists, and other professionals. Second, it is necessary to follow-up the undeniable cases of other genetic mutations, even if *BRCA1/2* is not mutated, and to lead them to genetic counseling and multigene panel testing. Individuals at risk of hereditary tumor syndromes should be informed about additional testing, including multigene panel testing. If there are no *BRCA1/2* mutations, there should be collaboration with the genetic medicine department to perform the test for those who request for it.

1.3 Risk Reduction Surgery

Contralateral risk-reducing mastectomy (CRRM) has been reported to reduce the risk of contralateral breast cancer [20–23]. Some studies have reported an improvement in overall survival (OS) [20, 21], while others have reported no effect [23, 24]. Heemskerk-Gerritsen BA et al. reported a prospective comparative study of 242 patients who received CRRM and 341 who did not receive CRRM but did receive surveillance (median follow-up 11.4 years after breast cancer diagnosis) and found that the CRRM group had a lower mortality rate than the surveillance group with a risk-reducing salpingo-oophorectomy (RRSO)-adjusted hazard ratio (HR): 0.49 (95% CI: 0.29–0.82) [21]. In a study of 698 breast cancer patients with *BRCA1/2* mutations (105 of whom underwent CRRM), Evans et al. reported an improvement in survival in the CRRM group (HR 0.37, $P = 0.008$) with a median follow-up period of 9 years [20]. In contrast, Brekelmans et al. reported that CRRM did not improve the OS of 170 breast cancer patients with *BRCA1* mutations [24]. Further investigation is required to determine the impact of CRRM on OS.

In 2019, Xiao et al. reported a meta-analysis that RRSO lowered the risk of developing breast cancer in previously diagnosed breast cancer patients (*BRCA1* HR: 0.51, 95% CI: 0.20–0.83 and *BRCA2* HR:0.24, 95% Cl: 0.05–0.52). Xiao et al. reported that RRSO could improve the OS of women with breast cancer (HR = 0.33, 95% Cl: 0.28–0.38) [25]. The NCCN guideline states that the choice of CRRM for women should be discussed on a case-by-case basis, and RRSO should be performed after completion of childbearing [14].

The number of prophylactic surgeries is expected to increase in the future, as *BRCA* genetic test is covered by health insurance in Japan. Several issues were also considered. First, the use of operating rooms is limited. Even if *BRCA1/2* mutations are identified, prophylactic surgery cannot be performed at the same time as breast cancer surgery due to the lack of an operating room. Second, it is necessary to cooperate with other departments such as plastic surgery, obstetrics, and gynecology. Additional surgical options for primary breast cancer surgery include nipple sparing and breast reconstruction (Fig. 2). However, it is often difficult to coordinate the surgery time and schedules of various departments, and surgeries must be performed on separate days. Smooth and prompt collaboration among various departments and a flexible operating room management system are important to start the treatment of primary breast cancer without delay and to allow patients choose a surgical procedure that satisfies their needs.

1.4 Surveillance

Regarding surveillance of breast cancer, the NCCN guidelines recommend annual mammograms with consideration of tomosynthesis and breast MRI surveillance with contrast agent in those who are treated for breast cancer and have not had

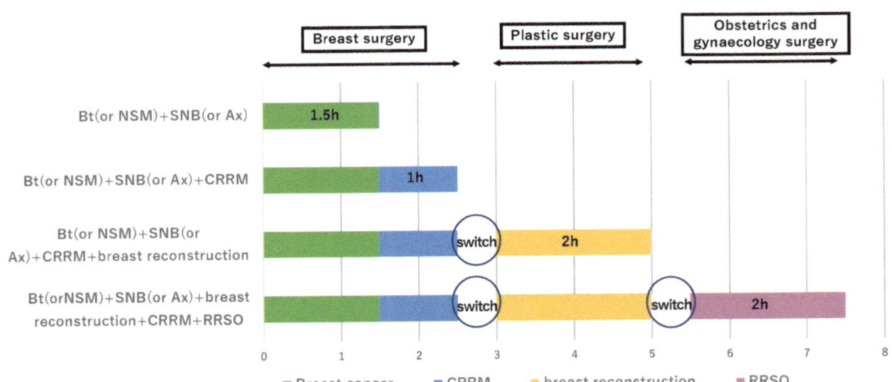

Fig. 2 An example of surgical options and time schedule for HBOC. Since there are various options and the operating time varies accordingly, cooperation with other departments and sections is essential for smooth operation

CRRM [14]. Previous reports have shown that surveillance with MRI has a higher sensitivity than conventional surveillance with mammography [26–28]. The HBOC clinical practice guidelines in Japan included a systematic review of three observational studies comparing survival rates for surveillance with and without breast MRI. Systematic reviews reported higher survival rates in the surveillance group with MRI than in the group without MRI (10-year survival 95.3 vs 87.7, 100 vs. 85.5, 3-year survival 100 vs. 92). However, the number of cases was small, and the observation period was short. Thus, further studies are needed to evaluate the extent of improvement in survival [29–32]. Regarding surveillance of ovarian cancer, there are some studies on the sensitivity and specificity of screening, including transvaginal ultrasound and CA-125. However, there is no evidence that these screenings are substitutes for RRSO screening. In the case of male breast cancer, screening for pancreatic and prostate cancer is required. *BRCA1/2* mutation-positive men have been reported to develop prostate cancer at a younger age more frequently than the general population, with poor clinicopathological features, treatment outcomes, and prognosis [6, 33, 34]. Regarding surveillance of prostate cancer, the NCCN and HBOC practice guidelines recommend surveillance by PSA starting at age 40 years. Regarding surveillance of pancreatic cancer, ultrasound endoscopy and contrast agent MRI are recommended. The NCCN guidelines state that pancreatic cancer surveillance may contribute to down staging; however, long-term studies are needed to determine whether down staging leads to improved survival, and no clear screening method has yet been proposed. Regarding the surveillance of melanoma, there are no specific screening guidelines, but general melanoma risk management, such as annual full-body skin examination and minimizing UV exposure, is recommended [14].

There are some future issues with surveillance. First, facilities for MRI and MR-guided biopsy are limited. In Japan, there are only 11 facilities where MR-guided biopsy is covered by health insurance (Jun 2022). In the future, it is necessary to develop facilities and train personnel to perform MR-guided biopsy. Second, screening for prostate and pancreatic cancers has not yet been established. Third, surveillance is not covered by the health insurance for *BRCA* mutation carriers who have not developed breast cancer.

1.5 PARP Inhibitor

PARP is an enzyme that repairs DNA single-strand breaks by repairing base breaks. When PARP is inhibited, single-strand DNA breaks accumulate, leading to double-strand DNA breaks at the replication fork. Normally, these breaks are repaired by the homologous recombination double-stranded DNA repair pathway of the tumor suppressor proteins such as *BRCA1* and *BRCA2* [35, 36]. In cells with *BRCA* pathological variants, DNA is not repaired, and PARP inhibitors induce cell death. A previous study (OlympiAD Clinical Trial) showed a significant increase in progression-free survival in the PARP inhibitor group compared with the standard therapy group in patients with inoperable or recurrent breast cancer with *BRCA1/2* mutations and HER2-negative disease who had received prior chemotherapy (HR 0.58, 95% Cl 0.43–0.80). The study did not show a significant increase in OS, but there was a trend toward improved OS [37]. In Japan, the use of PARP inhibitors in these patients was covered by health insurance in 2018.

A phase III trial (Olympia Clinical Trial) on the use of PARP inhibitors in patients with early stage breast cancer is also underway. The participants were high-risk early stage breast cancer patients who have germline *BRCA1/2* mutations, HER2-negative, and also received chemotherapy either preoperatively or postoperatively. The study showed a significantly longer invasive disease-free survival in the PARP inhibitor group than in the placebo group (HR 0.58, $P < 0.001$). Also, there was an improvement in OS (HR 0.68, $P = 0.02$) [38]. In the United States, the use of PARP inhibitors for these patients was approved by the Food and Drug Administration in March 2022. It is possible that these drugs will be covered by health insurance in Japan in the future.

1.6 Chemoprevention

The use of selective estrogen receptor modulators, such as tamoxifen and raloxifene, reduces the risk of invasive breast cancer in postmenopausal women considered at high risk of breast cancer [39–41]. One of the largest randomized trials on chemoprevention is the NSABP P-1 trial. The trial of 13,388 women with nonbreast cancer showed a reduction in the cumulative incidence of invasive breast cancer in the oral tamoxifen at 20 mg/day for the 5-year group compared with the placebo group after 7 years of follow-up (RR = 0.57, 95% CI = 0.46 to 0.70) [42]. However, tamoxifen toxicity, such as venous thromboembolic, endometrial cancer, and menopausal symptoms, is problematic. It is reported that low-dose tamoxifen at 5 mg/day for 3 years reduces the risk of local and contralateral recurrence with a limited toxicity in breast intraepithelial neoplasia, such as atypical ductal hyperplasia (ADH), noninvasive ductal carcinoma of the breast (DCIS), and noninvasive lobular carcinoma (LCIS) (HR, 0.48; 95% CI, 0.26 to 0.92; $P = 0.02$) [43]. Regarding *BRCA1/2* mutation carriers, there is a meta-analysis study that shows that tamoxifen reduces the incidence of contralateral breast cancer among *BRCA1/2* mutation carriers (*BRCA1/2*:RR = 0.56, 95% CI = 0.41–0.76, *BRCA1*:RR = 0.47, 95% CI = 0.37–0.60, *BRCA2*:RR = 0.39, 95% CI = 0.28–0.54) [44]. A study is currently underway in Japan to evaluate the efficacy of oral tamoxifen prophylaxis in *BRCA2* mutation carriers who have not developed breast cancer. Further accumulation of data on the preventive effects of tamoxifen is necessary in the future.

1.7 Beyond HBOC

Established breast cancer predisposition genes include *ATM*, *BARD1*, *CDH1*, *CHEK2*, *NF1*, *PALB2*, *PTEN*, *RAD51C*, *RAD51D*, and *TP53* other than *BRCA1/2*. In a population-based case-control study of 32,247 women with breast cancer (case patients) and 32,544 unaffected women (controls) in the United States, the prevalence of pathogenic variants in 12 established breast cancer–predisposition genes (*ATM*, *BARD1*, *BRCA1*, *BRCA2*, *CDH1*, *CHEK2*, *NF1*, *PALB2*, *PTEN*, *RAD51C*, *RAD51D*, and *TP53*) was 5.03% (95% CI, 4.79–5.27) among case patients and 1.63% (95% CI, 1.50–1.78) among controls. Among the case patients, the prevalence of pathogenic variants was high for *BRCA2* (1.29%; 95% CI, 1.18–1.42), *CHEK2* (1.08%; 95% CI, 0.98–1.20), and *BRCA1* (0.85%; 95% CI, 0.76–0.96). The breast cancer risk of pathogenic variants in predisposition genes was high for *BRCA1* (odds ratio, 7.62; 95% CI, 5.33–11.27), *BRCA2* (odds ratio, 5.23; 95% CI, 4.09–6.77), and *PALB2* (odds ratio, 3.83%; 95% CI, 2.68–5.63) [45]. Momozawa et al. reported the prevalence of pathogenic variants in 404 (5.7%) breast cancer cases and 67 (0.6%) controls in a study of 7051 breast cancer cases and 11,241 controls in Japanese. Among the case patients, the prevalence of pathogenic variants was high for *BRCA2* (2.71%), *BRCA1* (1.45%), *PALB2* (0.40%), and *CHEK2* (0.37%). The breast cancer risk of pathogenic variants in predisposition genes was high for *BRCA1* (odds ratio 33.0; 95% CI, 13.7–28.0), *PTEN* (odds ratio 17.6; 95% Cl, 2.6–753.3), *BRCA2* (odds ratio 16.4; 95% CI, 13.7–103.8), and *PALB2* (odds ratio 9.0; 95% CI, 3.4–29.7) [46]. Therefore, even if *BRCA1/2* expression is negative, other genetic mutations may be present. Surveillance should be conducted with genetic mutations in moderate breast cancer risk genes, such as *PALB2* and *CHECK2*. In addition, surveillance should be conducted for genetic mutations that significantly affect treatment and surveillance, such as Li-Fraumeni. The choice of cases to proceed with panel testing will be an issue for future consideration; however, *BRCA1/2* genetic test is covered by health insurance in Japan.

1.8 Genetic Test for Family

The NCCN guidelines recommend educating families and providing information about available resources and breast cancer risks. In Japan, a family member with *BRCA1/2* mutation career can undergo *BRCA1/2* genetic testing for the presence of specific mutations, but this is not covered by health insurance. It is reported that an annual incidence of breast cancer is 6.6% in unaffected *BRCA1/2* mutation careers according to the report of JOHBOC. Surveillance is important for *BRCA1/2* mutation carriers who have not yet developed breast cancer; however, such examinations are not covered by the health insurance. Breast MRI is expensive, even though it is particularly effective in detecting early breast cancer. The establishment of a surveillance system for *BRCA1/2* mutation carriers who do not develop breast cancer will be an issue for future consideration. In Japan, there are concerns such as insufficient information not offered to family members due to the nuclear family structure and accessibility to genetic medicine. It is expected that appropriate information will be provided in the future, for example, through the use of online medical services.

2 Conclusion

The identification of HBOC is highly effective and beneficial to the patient, as it expands the scope of treatment and follow-up. Therefore, it is important for physicians to provide medical care for HBOC.

References

1. Comprehensive molecular portraits of human breast tumours. Nature. 2012;490:61–70.
2. Tung N, Lin NU, Kidd J, et al. Frequency of germline mutations in 25 cancer susceptibility genes in a sequential series of patients with breast cancer. J Clin Oncol. 2016;34(13):1460–8.
3. Yoshida K, Miki Y. Role of BRCA1 and BRCA2 as regulators of DNA repair, transcription, and cell cycle in response to DNA damage. Cancer Sci. 2004;95(11):866–71.
4. Chen S, Parmigiani G. Meta-analysis of BRCA1 and BRCA2 penetrance. J Clin Oncol. 2007;25(11):1329–33.
5. Kuchenbaecker KB, Hopper JL, Barnes DR, et al. Risks of breast, ovarian, and contralateral breast cancer for BRCA1 and BRCA2 mutation carriers. JAMA. 2017;317(23):2402–16.
6. Nyberg T, Frost D, Barrowdale D, et al. Prostate cancer risks for male BRCA1 and BRCA2 mutation carriers: a prospective cohort study. Eur Urol. 2020;77(1):24–35.
7. Goggins M, Schutte M, Lu J, et al. Germline BRCA2 gene mutations in patients with apparently sporadic pancreatic carcinomas. Cancer Res. 1996;56(23):5360–4.
8. Iqbal J, Ragone A, Lubinski J, et al. The incidence of pancreatic cancer in BRCA1 and BRCA2 mutation carriers. Br J Cancer. 2012;107(12):2005–9.
9. Miki Y, Swensen J, Shattuck-Eidens D, et al. A strong candidate for the breast and ovarian cancer susceptibility gene BRCA1. Science. 1994;266(5182):66–71.
10. Wooster R, Bignell G, Lancaster J, et al. Identification of the breast cancer susceptibility gene BRCA2. Nature. 1995;378(6559):789–92.
11. Cropper C, Woodson A, Arun B, et al. Evaluating the NCCN clinical criteria for recommending BRCA1 and BRCA2 genetic testing in patients with breast cancer. J Natl Compr Cancer Netw. 2017;15(6):797–803.
12. George A. UK BRCA mutation testing in patients with ovarian cancer. Br J Cancer. 2015;113 Suppl 1(Suppl 1):S17–21.
13. Forbes C, Fayter D, de Kock S, Quek RG. A systematic review of international guidelines and recommendations for the genetic screening, diagnosis, genetic counseling, and treatment of BRCA-mutated breast cancer. Cancer Manag Res. 2019;11:2321–37.
14. National Comprehensive Cancer Netwok Clinical Practice Guidelines in Oncology. Genetic/Familial High-Risk Assessment: Breast, Ovarian, and Pancreatic. Version 2.2022 — March 9, 2022.
15. Yadav S, Hu C, Hart SN, et al. Evaluation of germline genetic testing criteria in a hospital-based series of women with breast cancer. J Clin Oncol. 2020;38(13):1409–18.

16. ESMO Prevention and Screening in BRCA Mutation Carriers and Other Breast/ovarian Hereditary Cancer Syndromes Guideline. http://interactiveguidelines.esmo.org/esmo-web-app/gl_toc/index.php?GL_id=53. Accessed 22 May 2022
17. Guidebook for Diagnosis and Treatment of Hreditary Breast and Ovarian Cancer Syndrome; 2017. http://johboc.jp/guidebook2017/toc/2-1index/cq1/. Accessed 29 May 2022
18. Guo F, Scholl M, Fuchs EL, Berenson AB, Kuo YF. BRCA testing and testing results among women 18-65 years old. Prev Med Rep. 2022;26:101738.
19. Sun L, Cui B, Wei X, et al. Cost-effectiveness of genetic testing for all women diagnosed with breast cancer in China. Cancers (Basel). 2022;14(7)
20. Evans DG, Ingham SL, Baildam A, et al. Contralateral mastectomy improves survival in women with BRCA1/2-associated breast cancer. Breast Cancer Res Treat. 2013;140(1):135–42.
21. Heemskerk-Gerritsen BA, Rookus MA, Aalfs CM, et al. Improved overall survival after contralateral risk-reducing mastectomy in BRCA1/2 mutation carriers with a history of unilateral breast cancer: a prospective analysis. Int J Cancer. 2015;136(3):668–77.
22. Metcalfe K, Gershman S, Ghadirian P, et al. Contralateral mastectomy and survival after breast cancer in carriers of BRCA1 and BRCA2 mutations: retrospective analysis. BMJ. 2014;348:g226.
23. van Sprundel TC, Schmidt MK, Rookus MA, et al. Risk reduction of contralateral breast cancer and survival after contralateral prophylactic mastectomy in BRCA1 or BRCA2 mutation carriers. Br J Cancer. 2005;93(3):287–92.
24. Brekelmans CT, Seynaeve C, Menke-Pluymers M, et al. Survival and prognostic factors in BRCA1-associated breast cancer. Ann Oncol. 2006;17(3):391–400.
25. Xiao YL, Wang K, Liu Q, Li J, Zhang X, Li HY. Risk reduction and survival benefit of risk-reducing Salpingo-oophorectomy in hereditary breast cancer: meta-analysis and systematic review. Clin Breast Cancer. 2019;19(1):e48–65.
26. Guindalini RSC, Zheng Y, Abe H, et al. Intensive surveillance with biannual dynamic contrast-enhanced magnetic resonance imaging downstages breast cancer in BRCA1 mutation carriers. Clin Cancer Res. 2019;25(6):1786–94.
27. Riedl CC, Luft N, Bernhart C, et al. Triple-modality screening trial for familial breast cancer underlines the importance of magnetic resonance imaging and questions the role of mammography and ultrasound regardless of patient mutation status, age, and breast density. J Clin Oncol. 2015;33(10):1128–35.
28. Warner E, Plewes DB, Hill KA, et al. Surveillance of BRCA1 and BRCA2 mutation carriers with magnetic resonance imaging, ultrasound, mammography, and clinical breast examination. JAMA. 2004;292(11):1317–25.
29. Evans DG, Kesavan N, Lim Y, et al. MRI breast screening in high-risk women: cancer detection and survival analysis. Breast Cancer Res Treat. 2014;145(3):663–72.
30. Chéreau E, Uzan C, Balleyguier C, et al. Characteristics, treatment, and outcome of breast cancers diagnosed in BRCA1 and BRCA2 gene mutation carriers in intensive screening programs including magnetic resonance imaging. Clin Breast Cancer. 2010;10(2):113–8.
31. Evans DG, Harkness EF, Howell A, et al. Intensive breast screening in BRCA2 mutation carriers is associated with reduced breast cancer specific and all cause mortality. Hered Cancer Clin Pract. 2016;14:8.
32. Japanese Organization of Heredity Breast and Ovarian Cancer. Guidelines for Diagnosis and Treatment of Hreditary Breast and Ovarian Cancer; 2021.
33. Castro E, Goh C, Olmos D, et al. Germline BRCA mutations are associated with higher risk of nodal involvement, distant metastasis, and poor survival outcomes in prostate cancer. J Clin Oncol. 2013;31(14):1748–57.
34. Na R, Zheng SL, Han M, et al. Germline mutations in ATM and BRCA1/2 distinguish risk for lethal and indolent prostate cancer and are associated with early age at death. Eur Urol. 2017;71(5):740–7.
35. Fong PC, Boss DS, Yap TA, et al. Inhibition of poly(ADP-ribose) polymerase in tumors from BRCA mutation carriers. N Engl J Med. 2009;361(2):123–34.

36. McCabe N, Turner NC, Lord CJ, et al. Deficiency in the repair of DNA damage by homologous recombination and sensitivity to poly(ADP-ribose) polymerase inhibition. Cancer Res. 2006;66(16):8109–15.
37. Robson M, Im SA, Senkus E, et al. Olaparib for metastatic breast cancer in patients with a germline BRCA mutation. N Engl J Med. 2017;377(6):523–33.
38. Tutt ANJ, Garber JE, Kaufman B, et al. Adjuvant olaparib for patients with BRCA1- or BRCA2-mutated breast cancer. N Engl J Med. 2021;384(25):2394–405.
39. Cummings SR, Eckert S, Krueger KA, et al. The effect of raloxifene on risk of breast cancer in postmenopausal women: results from the MORE randomized trial. Multiple Outcomes of Raloxifene Evaluation. JAMA. 1999;281(23):2189–97.
40. Lippman ME, Cummings SR, Disch DP, et al. Effect of raloxifene on the incidence of invasive breast cancer in postmenopausal women with osteoporosis categorized by breast cancer risk. Clin Cancer Res. 2006;12(17):5242–7.
41. Cuzick J, Sestak I, Bonanni B, et al. Selective oestrogen receptor modulators in prevention of breast cancer: an updated meta-analysis of individual participant data. Lancet. 2013;381(9880):1827–34.
42. Fisher B, Costantino JP, Wickerham DL, et al. Tamoxifen for the prevention of breast cancer: current status of the National Surgical Adjuvant Breast and Bowel Project P-1 study. J Natl Cancer Inst. 2005;97(22):1652–62.
43. DeCensi A, Puntoni M, Guerrieri-Gonzaga A, et al. Randomized placebo controlled trial of low-dose tamoxifen to prevent local and contralateral recurrence in breast intraepithelial neoplasia. J Clin Oncol. 2019;37(19):1629–37.
44. Xu L, Zhao Y, Chen Z, Wang Y, Chen L, Wang S. Tamoxifen and risk of contralateral breast cancer among women with inherited mutations in BRCA1 and BRCA2: a meta-analysis. Breast Cancer. 2015;22(4):327–34.
45. Hu C, Hart SN, Gnanaolivu R, et al. A population-based study of genes previously implicated in breast cancer. N Engl J Med. 2021;384(5):440–51.
46. Momozawa Y, Iwasaki Y, Parsons MT, et al. Germline pathogenic variants of 11 breast cancer genes in 7,051 Japanese patients and 11,241 controls. Nat Commun. 2018;9(1):4083.

Guidelines for Diagnosis and Treatment of HBOC: Methods and Products

Atsuko Kitano

Abstract This guideline was developed to support shared decision-making between individuals diagnosed with hereditary breast and ovarian cancer (HBOC) and health care providers regarding treatment and surveillance methods after diagnosis.

The values of genetic diseases are highly individualized, and in HBOC practice, "prophylactic removal" of healthy organs may be considered from the perspective of preventing the onset of the disease. Therefore, shared decision-making between the patient and the health care provider that takes into account diverse values is extremely important.

This guideline was prepared in compliance with the preparation method of "The Minds Clinical Guideline Preparation Manual 2017". Physicians, nurses, genetic counselors, and patients from various disciplines involved in HBOC practice participated in the development of this clinical guideline. After conducting a high-quality systematic review, recommendations were developed from multiple perspectives. This chapter describes the methodology used to develop these guidelines.

Keywords HBOC · Clinical guideline · Shared decision-making

1 Purpose of the Guidelines

This guideline was developed to support shared decision-making between individuals diagnosed with hereditary breast and ovarian cancer syndrome (HBOC) (*BRCA* pathological variant carriers) and health care providers regarding post-diagnosis treatment and surveillance methods. The project was also designed to support shared decision-making between HBOC and diagnosed parties (*BRCA* pathological variant holders) and medical care providers regarding post-diagnosis treatment and

A. Kitano (✉)
Department of Medical Oncology, St. Luke's International Hospital, Tokyo, Japan
e-mail: atsukita@luke.ac.jp

D. Aoki et al. (eds.), *Practical Guide to Hereditary Breast and Ovarian Cancer*,
https://doi.org/10.1007/978-981-99-5231-1_2

surveillance. In addition, due to the unique nature of the genetic disease, the impact of the disease on the blood relatives of the parties involved is not small. This guideline is also intended to support shared decision-making with the relatives of the patient regarding their choice to be informed, their choice to undergo genetic testing, and the impact of the test results.

2 Significance of Guideline Creation

Hereditary breast and ovarian cancer syndrome (HBOC syndrome, hereafter referred to as HBOC) is a cancer susceptibility syndrome including breast and ovarian cancer caused by germline mutations in *BRCA*. In Japan, *BRCA* genetic testing was covered by insurance in 2018 for ovarian and breast cancer patients as a companion diagnosis for therapeutic drug selection. In April 2020, *BRCA* genetic testing was extended to ovarian cancer patients and some breast cancer patients. In addition, risk-reducing surgery and surveillance are also covered by insurance for breast and ovarian cancer patients with *BRCA1/2* pathological variants.

In recent years, genetic medicine has come to be utilized in actual clinical practice, and not only germ-line mutations but also somatic mutations are now covered by insurance, and analysis from both directions has led to the shift from conventional treatment based on the primary site (e.g. breast cancer and lung cancer) to personalized treatment for the genes expressed in the cancer. This dual analysis has led to a shift from conventional treatment based on the primary site (breast cancer, lung cancer, etc.) to individualized treatment for the genes expressed in the cancer. In this trend, it is also necessary to deal with HBOC that are found as secondary findings. In particular, due to the unique nature of HBOC as a genetic disease, it affects not only the patient but also his/her blood relatives.

In various choices, guidelines to support shared decision-making by the parties concerned (*BRCA* pathological variant holders) and medical professionals, while respecting the values and individuality of the parties concerned, and involving them as a team, seem to be indispensable.

In Japan, a research group entitled "Study on the Elucidation of Clinical Genetic Characteristics of Hereditary Breast and Ovarian Cancer in Japan and Improvement of Life Outcomes Using Genetic Information" (principal investigator: Masami Arai) was organized under the Health and Labour Sciences Research (Comprehensive Research Project for the Promotion of Cancer Control). The research group's research results were the "Guide to Hereditary Breast and Ovarian Cancer Syndrome (HBOC), 2017 Edition. The 2021 version of the guideline will adhere to the Minds "Practice Guideline Development Manual 2017" and will reflect the diverse values of *BRCA* variant holders and health care providers, and aim to be a guideline that can be used in decision-making.

3 Method of Preparation of These Guidelines

3.1 Preparation Members

Three committee worked independently on this guideline: the "Supervisory Committee Member" who oversees the overall guideline, the "Guideline Development Committee Member" who drafts recommendations, and the "Systematic Review Committee Member" who conducts systematic reviews.

In addition to physicians, members included genetic nurses, certified genetic counselors, and statisticians. Specialty, gender, and geographic region were considered in the selection of the individuals.

3.2 Establishment of Clinical Queries

In developing this guideline, the supervising committee members and guideline development committee members created a clinical algorithm for this area. Based on the practice algorithm, key clinical questions were set. For clinical questions that were not designated as BQs, PICO-style Clinical Questions (CQs) were created and a literature search was conducted. CQs for which the evidence was deemed insufficient to evaluate the total body of evidence as a CQ were designated as Future Questions (hereinafter referred to as FQs).

3.3 Systematic Review and Evidence Synthesis

For each CQ, an evidence assessment sheet was used to evaluate the evidence for each outcome. The evidence evaluation sheet for each outcome assessed the risk of bias (selection bias, execution bias, detection bias, case attrition bias, etc.), ascendancy, non-directness, non-consistency, imprecision, and publication bias. After the evidence evaluation for each outcome was completed, an evidence rating sheet for the total body of evidence was used to evaluate the total body of evidence for the entire CQ. On the evaluation sheet for the evidence synthesis, each CQ outcome was rated for risk of bias (selection bias, execution bias, detection bias, case attrition bias, etc.), inconsistency, imprecision, and publication bias to determine the strength of the evidence.

After generating the total body of evidence, a qualitative systematic review was conducted and SR reports were generated for each CQ.

The entire process from evidence selection to SR report generation was performed independently by the SR committee members and did not involve the guideline development team.

3.4 Recommendation-Making

The recommendation-making meeting was held independently of the SR committee members and was attended by five physicians from the supervisory committee and three physicians from each area leader among the guideline development committee members. Also participating in the meeting were, in addition to the physicians, one nurse specialist in genetic nursing and one certified genetic counselor, and three representatives of the parties concerned also participated.

At the recommendation-making meeting, for each CQ to be discussed, the financial and academic conflicts of interest of the voters were disclosed, and committee members with conflicts of interest participated only in the discussion and did not vote.

The EtD frameworks of the GRADE system were used to create the recommendations. The EtD frameworks use the following nine criteria to comprehensively evaluate the CQs from various perspectives.

Criterion 1. Priority of the problem.
Criterion 2. Desirable effects.
Criterion 3. Undesirable effects.
Criterion 4. Certainty of evidence.
Criterion 5. Values.
Criterion 6. Balance of effects.
Criterion 7. Cost-effectiveness.
Criterion 8. Acceptability.
Criterion 9. Feasibility.

Prior to the recommendation decision meeting, participants evaluated each CQ on their own using the EtD frameworks, and the editorial board members summarized the pre-voting results and used them as materials for the recommendation decision meeting. At the recommendation decision meeting, all participants discussed and re-voted for each decision in the EtD frameworks. The final "type of recommendation" was decided. The content of the recommendation decision meeting was clearly stated in the main text of the guideline, in an attempt to make the process of recommendation decision transparent.

All the meetings were held online, and the recorded recordings of the meetings were used as reference materials when writing the recommendations and explanatory notes.

4 Type of Recommendation

Strong recommendation against the intervention in question.

Conditional recommendation against the intervention.

Conditional recommendation for either the intervention or the comparison.

Conditional recommendation for the intervention.

Strong recommendation for the intervention.

Conditional recommendation for either the intervention or the control" can only be selected if both the control and the intervention are recommended.

5 External Evaluation

We received external evaluations from external evaluation committee members and related societies. Public comments were also solicited on the JSCE website. Public comments were solicited using the public comment solicitation support service provided by Minds.

The sections pointed out in the external evaluation were added and revised in the final version.

6 Summary

This guideline was developed to support shared decision-making between HBOC patients and medical professionals regarding post-diagnosis treatment and surveillance methods. In addition to physicians, various medical professionals were involved in the development of this document, and the parties were invited to participate in the decision-making process for recommendations.

The creation of the study was done in full compliance with the Minds 2017 edition, which resulted in the creation of a high-quality evidence synthesis. In addition, the EtD framework was used to make recommendations based on a wide range of criteria.

We hope that this guideline will reach not only all medical professionals, but also HBOC patients and their families, and that HBOC treatment will be promoted based on satisfactory shared decision-making.

HBOC from a Plastic Surgeon's Perspective

Kenta Tanakura

Abstract The year 2020 was a year of great change in hereditary breast and ovarian cancer (HBOC) treatment in Japan, as the 2020 revision made testing, diagnosis, and risk-reducing surgery covered by public insurance. In Japan, the treatment, which has been performed in a limited number of facilities and on a limited number of patients, will become widely available. In breast cancer treatment, breast reconstruction is performed as part of the treatment to facilitate acceptance of mastectomy, which is the core of treatment. In the practice of HBOC, which also involves the healthy breast, the significance is even greater. The purpose of this article is to share our knowledge of breast reconstruction, including our experience in self-funded treatment.

Keywords Hereditary breast and ovarian cancer · Risk-reducing mastectomy · Breast reconstruction · Breast implant · Deep inferior epigastric artery perforator flap · Profunda femoris artery perforator flap · Fat graft · Life style · Life stage

1 Breast Reconstruction in Japan

Breast reconstruction in Japan has been a self-funded procedure until prosthetic reconstruction was covered by insurance. 2013 saw the introduction of tissue expanders (TE) and breast implants (BI). The number of cases of one-stage reconstruction with BI has increased to about 5000 in 2018 [1] (Fig. 1). The number of cases of primary first-stage reconstruction with BI was limited to nipple-sparing mastectomy and subpectoral reconstruction and remained at around 1000 cases per year. Together with autologous tissue reconstruction, breast reconstruction was taking root in Japan; the 2018 breast cancer practice guidelines issued by the Japanese

---

K. Tanakura (✉)
Department of Plastic and Reconstructive Surgery/Breast Center, Mitsui Memorial Hospital, Tokyo, Japan
e-mail: tanakura-kenta@mitsuihosp.or.jp

D. Aoki et al. (eds.), *Practical Guide to Hereditary Breast and Ovarian Cancer*,
https://doi.org/10.1007/978-981-99-5231-1_3

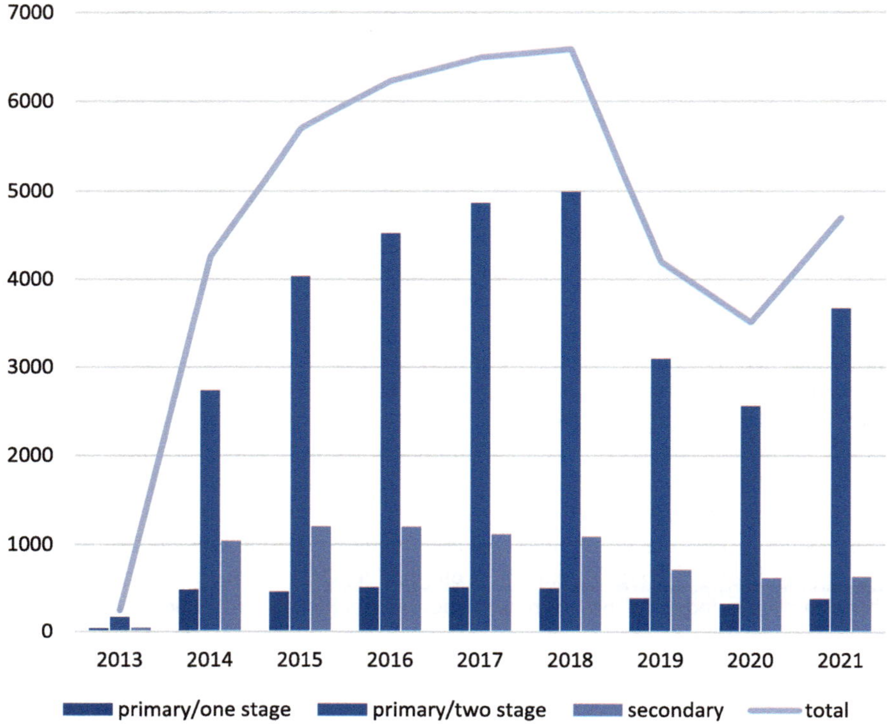

Fig. 1 Trends in breast reconstruction with breast implants in Japan, from insurance coverage in 2013 to 2021, are shown. The total is shown as a line graph, and the classification by surgical occasion is shown as a bar graph. Adapted from the annual report of the Japanese Oncoplastic Breast Surgery Society [1]

Breast Cancer Society stated, "Since almost all breast reconstruction procedures, including BI, are now covered by insurance, information must be provided to all patients who are eligible for such procedures." As for autologous tissue reconstruction, according to the NDB open data, although the number of facilities performing the procedure and the numbers of cases are unevenly distributed by region, the total number of cases is generally less than 1900 per year, including free flaps and pedicled flaps. In other words, about 8000 breast reconstruction procedures are performed annually in Japan, with breast implant reconstruction accounting for three-fourths and autologous tissue reconstruction for one-fourth.

We can see that the number of reconstructions using BI, which was insured in 2013, increased until 2018, but was halved by the "Allergan Crisis" in 2019, and has not returned to its previous number even after the launch of the replacement product.

2 RRM and Breast Reconstruction for HBOC in the Era of Self-Funded Treatment

In Japan, risk-reducing mastectomy (RRM) and risk-reducing bilateral salpingo-oophorectomy (RRSO) for HBOC have been covered by insurance since April 2020. As mentioned above, breast implant reconstruction, along with TEs, was covered by insurance in 2013, but because mixed treatment is not allowed by custom, patients who receive RRM at self-funded costs before 2020 were required to also receive breast reconstruction at self-funded costs. This made it extremely difficult for autologous tissue reconstruction, which is approximately 2–4 times more expensive than breast implant reconstruction even on a public insurance point basis, to become an option. The author also experienced several RRM reconstructions during this period, but all of them were breast implant reconstructions, and none of them were autologous tissue reconstructions. The author's first experience with breast reconstruction for an HBOC patient was at the Cancer Institute Hospital, where bilateral breast implant reconstruction was performed at her own expense for a *BRCA2*-positive woman in her 30s with heterochronic bilateral breast cancer [2]. The following figures (Figs. 2 and 3) are cases of breast reconstruction for HBOC patients during this period. These cases were two-stage reconstructions for synchronous bilateral breast cancer or contralateral risk-reducing mastectomy (CRRM), but some cases were performed with NSM and direct-to-implant to reduce hospitalization and cost.

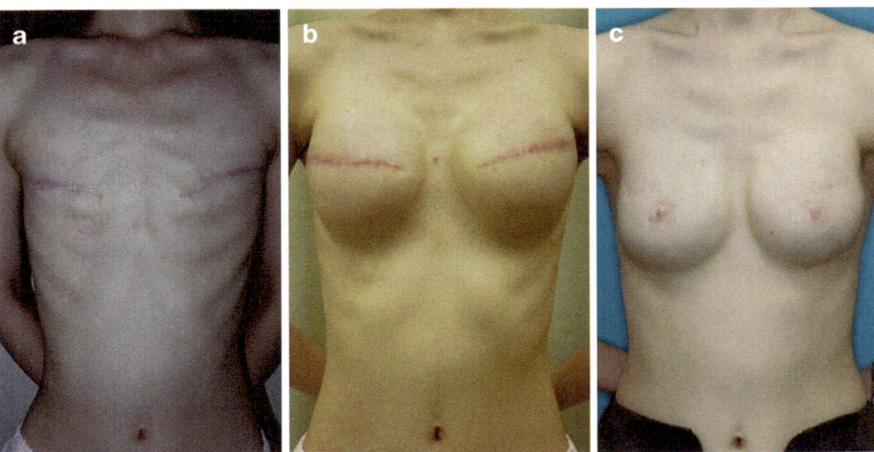

Fig. 2 Female in her late 20s, *BRCA1*-positive. After bilateral total mastectomy for synchronous bilateral breast cancer, (**a**) she requested reconstructive surgery at her own expense. TEs were inserted in the bilateral breasts (**b**) followed by anatomical BIs. She is now 8 years postoperatively with no particular complications (**c**)

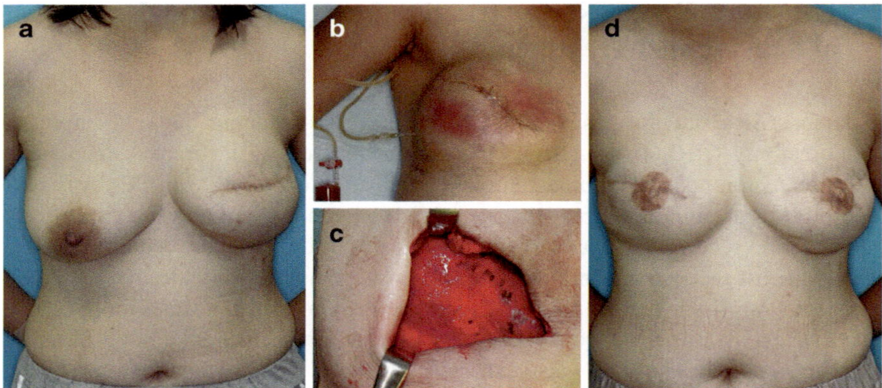

Fig. 3 A woman in her late 40s. She underwent a total mastectomy, sentinel node biopsy, and TE insertion for left breast cancer, followed by a BI replacement. (**a**) She subsequently tested positive for *BRCA1* and requested CRRM and reconstruction at her own expense. A similar total mastectomy and insertion of a TE was performed on the right side, but the patient developed an infection 2 weeks postoperatively. Ultrasound revealed a fluid retention, which was punctured and submitted to culture. (**b**) Immediately, antibiotics were empirically started, and surgery was performed. After thorough debridement, (**c**) it was determined that immediate replacement of the TE was possible. Subsequently, the patient underwent replacement with a BI. Five years after the surgery, there have been no complications (**d**)

3 BIA-ALCL and the "Allergan Crisis" in Japan

Since 2017, reports of breast implant associated anaplastic large cell lymphoma (BIA-ALCL) with a higher incidence in the macro-textured type with a larger surface area [3], and further reports in 2019 [4] have led to the suspension of the sale of this type of implant in many countries [5–7], and on July 24, 2019, Allergan's BIs and TEs with Biocell®, a macro-textured surface finish, were recalled worldwide [8]. The company was one of the three largest manufacturers in the United States and had some market share worldwide. In particular, in Japan, it was the only manufacturer approved for use under public insurance. The company's product line was also discontinued in Japan, which meant that not only could breast reconstruction with BIs no longer be performed in that country, but also primary reconstruction using TEs could no longer be performed there. By the end of the same year, the smooth type was approved and reconstruction could resume. In 2021, the number of implant cases recovered to about 4500, but this is far from the 6500 that existed before the uproar [1] (Fig. 1).

4 Properly Fearing BIA-ALCL

BIA-ALCL is a rare T-cell lymphoma. It occurs around textured-type breast implants and has a lifetime incidence of 1/2207–1/86029 [4, 9].

It typically develops with enlargement associated with delayed seroma of the breast containing a breast implant. Symptoms include delayed seroma in 80% and mass in 40% [9, 10]. By April 2021, 993 cases had been reported worldwide.

It is difficult for patients as well as physicians to understand sensitively the risk of this disease, which accounts for less than 0.1% of cases. The following is the information necessary to correctly fear this disease.

Radiation-induced sarcoma is a very rare complication (that breast surgeons believe) that occurs after irradiation, most commonly angiosarcoma, with an incidence of approximately 1/300 per decade [11]. Stewart-Treves syndrome, a vascular sarcoma associated with lymphedema, also has an incidence of 1/220–1/1400 at 10–15 years [12, 13]. Both have a higher incidence than BIA-ALCL, are more aggressive, and have a poorer prognosis. However, these diseases are not screened for.

Next, let us look at breast cancer itself. The current lifetime incidence rate of breast cancer in Japan is 1/9 [14]. It is a disease with a 200-fold higher risk of developing breast cancer than BIA-ALCL. Breast cancer is therefore a target for cancer screening in Japan with the aim of improving prognosis. According to the Japan Cancer Society, the detection rate of breast cancer by this single breast cancer screening is 0.24%. It is important to keep in mind that even for a disease with such a high lifetime morbidity, the screening positive rate is so low. BIA-ALCL is not a suitable screening target because the incidence is too low. Biennial imaging screening, currently advocated by the Japan Oncoplastic Breast Surgery Society (JOPBS), is also intended to search for damage, which occurs in about 1/9 of 10 years [15, 16].

Current recommendations for intervention in symptomatic patients are appropriate; algorithms by the NCCN and JOPBS have been proposed. For example, when approaching fluid retention and peri-implant masses, one should first keep in mind the exclusion of infection and local recurrence of breast cancer, which are by far more common than the rare disease of BIA-ALCL.

Also, total capsulectomy to prevent the development of BIA-ALCL is not recommended due to lack of evidence, as there have been cases of the disease occurring after the procedure was performed [17].

5 Breast Reconstruction After RRM After Insurance Coverage of RRM

The most significant feature of breast reconstruction for HBOC is that it is bilateral. Also, due to the fact that insurance coverage has only been available for a short time and cultural differences with Western countries, it will take time for patients to accept bilateral simultaneous excision. We also need to consider CRRM cases with anisochronous defects.

The convenience and minimally invasive nature of BI reconstruction makes itself more convenient in the sense that bilateral reconstruction is not affected by the morphology of the healthy side. The relatively young age of the patient also contributes to the advantage of BI when pregnancy is considered. However, recent quality of life studies based on patient-reported outcome measures, such as Breast-Q, have shown that autologous reconstruction is superior to BI reconstruction in long-term outcomes [18], and there is a possibility that the use of autologous reconstruction in younger patients will increase QALYs. In Japan, autologous tissue reconstruction for RRM has not been performed due to the burden of self-funded treatment, but insurance coverage has opened the way.

The most common autologous reconstruction is a skin flap using the lower abdomen, especially the deep inferior epigastric artery perforator flap (DIEP), which preserves the rectus abdominis muscle. The deep inferior epigastric artery, which feeds the rectus abdominis muscle, is usually used on one side for this flap, but it can be divided near the median line and the left and right sides can be harvested as separate skin flaps. Therefore, it can be used for bilateral reconstruction if both sides are reconstructed at the same time (Fig. 4). Conversely, it cannot be taken twice. In addition to concerns about pregnancy, the use of this procedure in young patients should be done with caution.

Another typical autologous reconstruction is the latissimus dorsi flap (LD), which is a pedicled flap from the back. This flap is capable of reconstructing moderately sized breasts and has been used in combination with fat grafting in recent years. The LD muscles exist bilaterally and can theoretically be used for bilateral

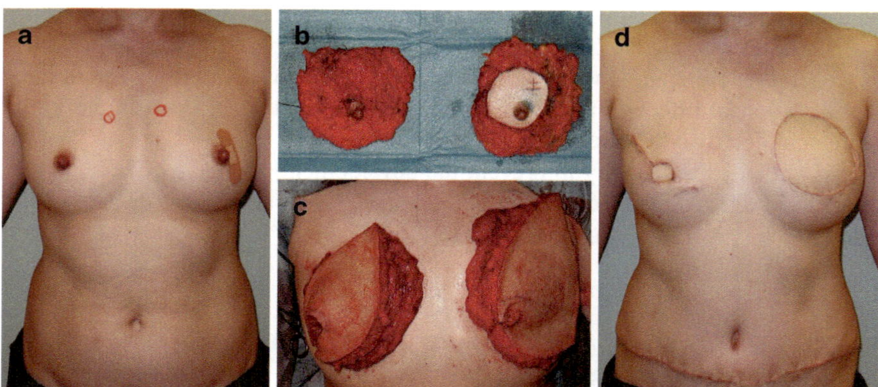

Fig. 4 Female in her early 40s, *BRCA2*-positive. She underwent a left total mastectomy, CRRM (skin-sparing mastectomy) on the right and immediate bilateral reconstruction with a DIEP flap. (**a**) The resection weight was 336 g on the left with lesion and 224 g on the right. (**b**) Subsequently, the left lesion was reconstructed with a 364-g flap and the right lesion with a 250-g flap. (**c**) The patient was discharged without complications on postoperative day 6. The picture postoperative 1 month. (**d**) 10 months later, an RRSO was performed laparoscopically

reconstruction. However, because harvesting of them always requires repositioning of the patient, the actual application may be limited considering the multiple repositionings during simultaneous bilateral reconstruction.

A skin flap that has recently attracted attention and is often selected by the author is Profunda femoris Artery Perforator flap (PAP), a relatively new skin flap whose application to breast reconstruction was reported by Allen et al. [19] in 2012. The skin flap is harvested from the medial posterior thigh and uses the skin perforator flap of the femoral deep artery running within the adductor magnus muscle as the vascular stem. In our own experience, the average volume for Japanese patients is about 240 g, making it suitable for reconstruction of breasts of moderate volume. The unique feature of this skin flap is that it can be taken separately from the left and right thighs. Therefore, it can be used for reconstruction of metachronous bilateral breast cancer or CRRM. In addition, there is no concern about pregnancy, since the donor site is not in the abdomen. The skin flap is suitable for reconstruction of autologous tissue in young patients. In addition, the skin flap can be used to reconstruct a large breast by taking skin flaps from both sides and using them to reconstruct one breast.

6 Fat Grafting

Fat grafting is a minimally invasive and scarless method of tissue transfer. Fat tissue is suctioned, purified in some way, and then injected into the desired area. Although fat grafting has been around for a long time, it was once shunned because of inconsistent results; however, after Coleman et al. proposed a refined method in the 1990s [20–22], it was reevaluated and widely used. In order to follow the current widely agreed-upon method, there is a limit to the amount of transplantation that can be performed in a single procedure in order to ensure safe and complication-free surgery. Therefore, applications include touch-up of a breast reconstructed with BI or autologous tissue, modification of breast-conserving surgery, and whole-breast reconstruction that gradually gains capacity over multiple procedures.

First, touch-up of the reconstructed breast [23–25] is to eliminate the left-right difference that remains as a result of BI or flap reconstruction. Step-off deformity is more likely to occur during BI reconstruction and relatively small perforator skin flaps other than DIEP (e.g., thigh, buttock, and lumbar) in Japanese, where subcutaneous fat is thin and mastectomy skin flaps are often thin after mastectomy. Mastectomy remains an unavoidable and cornerstone treatment even in this era of highly advanced drug therapy. Breast reconstruction is performed as part of breast cancer treatment to reduce or eliminate the sense of loss and ease the psychological burden, making treatment easier to receive and accept. If the goal of breast reconstruction is to rebuild the patient's mind and life, for example, by improving the deformity of the medial breast or axillary folds, the patient can wear a V-neck or camisole as she did before surgery. The ultimate goal of breast reconstruction surgery, the authors believe, is for the patient to forget that she had breast cancer. Fat grafting is an indispensable procedure for this purpose.

It can also be applied to correct deformities after breast conservation surgery [26]. Mildly depressed deformities are good candidates for fat grafting. In general, however, fat grafting for deformities after breast conserving surgery often requires the management of severe scarring. This is technically challenging and requires skill.

Total breast reconstruction with fat grafting is the closest to ideal reconstruction, but it is very challenging [27, 28]. Methods using TE, BI, and external negative pressure devices [29] have been reported. In any case, the breast after mastectomy is often thinnest in the area just below the nipple, which should be the projection top, and it is necessary to gradually increase the thickness of the tissue through multiple procedures. In most of the cases in our own experience, at least three cycles were required.

7 Lifestyle and Life Stage

Three types of reconstruction for RRM of HBOC are presented: BI, autologous tissue, and fat grafting. As mentioned above, there are reports of better QOL with autologous tissue reconstruction compared with BI reconstruction in the long term. Based on the treatment methods currently covered by insurance in Japan, it would be better for patients to undergo autologous tissue reconstruction at some point in their lives. However, it is necessary to take into account that the incidence of breast cancer in Japan is bimodal, with the age range of 45 and 65 years [30], and that many of the patients who are eligible for reconstruction are women in their 40s. They are very busy in both career and child-rearing aspects. The fact that hospitalization for autologous tissue reconstruction in Japan is generally two weeks, and that it still takes about four weeks postoperatively to return to physical labor, along with the fear of a new scar at the donor site, make it clear that autologous tissue reconstruction is not an option for all women. This is where the value of simple, minimally invasive BI reconstruction is found. On the other hand, in addition to low satisfaction, BI reconstruction has the inevitable complication of rupture.

Consider the case of a woman in her mid-40s who opted for BI reconstruction due to her busy schedule, only to discover 15 years later that her BI had ruptured. At this point in her life, she has finished raising her children and is heading toward retirement from her job. If she is forced to undergo reconstructive surgery at such a time, the previously unobtainable option of autologous reconstruction becomes a reality.

In other words, lifestyle limits the reconstructive options available. And even for the same patient, the options will change as the patient progresses through life stages. Breast reconstruction is a life-long necessity, and reconstructive surgeons are required to consider the patient's lifestyle and life stage and to provide breast reconstruction that is in harmony with the patient's life.

References

1. http://jopbs.umin.jp/medical/guideline/docs/gappeisho2021.pdf. Accessed 22 June 2022.
2. Tanakura K, Sawaizumi M, Yajima K, et al. Breast reconstruction for familial breast cancer with gene diagnosis. Jpn P R S. 2012;32:340–3.
3. Loch-wilkinson A, Beath KJ, Knight RJW, et al. Breast implant-associated anaplastic large cell lymphoma in Australia and New Zealand: high-surface-area textured implants are associated with increased risk. Plast Reconstr Surg. 2017;140(4):645–54.
4. Magnusson M, Beath K, Cooter R, et al. The epidemiology of breast implant associated large cell lymphoma in Australia and New Zealand confirms the highest risk for grade 4 surface breast implants. Plast Reconstr Surg. 2019;143:1285.
5. Risk of Breast Implant Associated-Anaplastic Large Cell Lymphoma (BIA-ALCL). Health Sciences Authority (Singapore). May 10, 2019.
6. Portant interdiction de mise sur le marché, de distribution, de publicité et d'utilisation d'implants mammaires à enveloppe macro-texturée et d'implants mammaires polyuréthane, ainsi que retrait de ces produits. ANSM. Apr 2, 2019.
7. Health Canada suspends Allergan's licences for its Biocell breast implants after safety review concludes an increased risk of cancer. Health Canada website. https://recalls-rappels.canada.ca/en/alert-recall/health-canada-suspends-allergan-s-licences-its-biocell-breast-implants-after-safety. Accessed 25 June 2022.
8. Allergan Voluntarily Recalls BIOCELL® Textured Breast Implants and Tissue Expanders. FDA website. https://www.fda.gov/safety/recalls-market-withdrawals-safety-alerts/allergan-voluntarily-recalls-biocellr-textured-breast-implants-and-tissue-expanders. Accessed 25 June 2022.
9. Clemens MW, McGuire PA. Discussion: a prospective approach to inform and treat 1340 patients at risk for BIA-ALCL. Plast Reconstr Surg. 2019;144(01):57–9.
10. Clemens MW, Medeiros LJ, Butler CE, et al. Complete surgical excision is essential for the management of patients with breast implant-associated anaplastic large-cell lymphoma. J Clin Oncol. 2016;34(2):160–8.
11. Yap J, et al. Sarcoma as a second malignancy after treatment for breast cancer. Int J Radiat Oncol Biol Phys. 2002;52:1231–7.
12. Fitzpatrick PJ. Lymphangiosarcoma and breast cancer. Can J Surg. 1969;12:172–7.
13. Shirger A. Postoperative lymphedema: etiologic and diagnostic factors. Med Clin North Am. 1962;46:1045–50.
14. https://ganjoho.jp/reg_stat/statistics/stat/summary.html. Accessed 22 June 2022.
15. Maxwell GP, Van Natta BW, Bengtson BP, Murphy DK. Ten-year results from the Natrelle 410 anatomical form-stable silicone breast implant core study. Aesthet Surg J. 2015;35(2):145–55.
16. Stevens WG, Calobrace MB, Alizadeh K, et al. Ten-year core study data for Sientra's food and drug administration–approved round and shaped breast implants with cohesive silicone gel. Plast Reconstr Surg. 2018;141:7S–19S.
17. Tanna N, Calobrace MB, Clemens MW, et al. Not all breast explants are equal: contemporary strategies in breast explantation surgery. Plast Reconstr Surg. 2021;147(4):808–18.
18. Eltahir Y, Krabbe-Timmerman IS, Sadok N, Werker PMN, de Bock GH. Outcome of quality of life for women undergoing autologous versus alloplastic breast reconstruction following mastectomy: a systematic review and meta-analysis. Plast Reconstr Surg. 2020;145(5):1109–23.
19. Allen RJ, Haddock NT, Ahn CY, Sadeghi A. Breast reconstruction with the profunda artery perforator flap. Plast Reconstr Surg. 2012;129(1):16e–23e.
20. Coleman SR. Lipoinfiltration of the upper lip white roll. Aesthet Surg J. 1994;14(4):231–4.
21. Coleman SR. The technique of periorbital lipoinfiltration. Oper Tech Plast Reconstr Surg. 1994;1:20–6.
22. Coleman SR. Structural fat grafting. St. Louis, MO: Quality Medical Publishing; 2004.
23. Spear SL, Wilson HB, Lockwood MD. Fat injection to correct contour deformities in the reconstructed breast. Plast Reconstr Surg. 2005;116(5):1300–5.

24. Kanchwala SK, Glatt BS, Conant EF, Bucky LP. Autologous fat grafting to the reconstructed breast: the management of acquired contour deformities. Plast Reconstr Surg. 2009;124(2):409–18.
25. Komorowska-Timek E, Turfe Z, Davis AT. Outcomes of prosthetic reconstruction of irradiated and nonirradiated breasts with fat grafting. Plast Reconstr Surg. 2017;139(1):1e–9e.
26. Delay E, Gosset J, Toussoun G, et al. Efficacy of lipomodelling for the management of sequelae of breast cancer conservative treatment. Ann Chir Plast Esthet. 2008;53(2):153–68.
27. Howes BH, Fosh B, Watson DI, et al. Autologous fat grafting for whole breast reconstruction. Plast Reconstr Surg Glob Open. 2014;2(3):e124. Published 2014 Apr 7
28. Longo B, Laporta R, Sorotos M, Pagnoni M, Gentilucci M, Santanelli di Pompeo F. Total breast reconstruction using autologous fat grafting following nipple-sparing mastectomy in irradiated and non-irradiated patients. Aesthet Plast Surg. 2014;38(6):1101–8.
29. Khouri RK, Rigotti G, Khouri RK Jr, et al. Tissue-engineered breast reconstruction with Brava-assisted fat grafting: a 7-year, 488-patient, multicenter experience. Plast Reconstr Surg. 2015;135(3):643–58.
30. Hayashi N, Kumamaru H, Isozumi U, et al. Annual report of the Japanese breast cancer registry for 2017. Breast Cancer. 2020;27:803–9.

Part II
Gynecology

Serous Tubal Intraepithelial Carcinoma (STIC) and Precancerous Lesions in Risk-Reducing Salpingo-oophorectomy (RRSO) Specimens

Kenta Masuda and Daisuke Aoki

Abstract Risk-reducing salpingo-oophorectomy (RRSO) is recommended for women with a *BRCA1* or *BRCA2* (*BRCA1/2*) pathogenic variant because there is no effective surveillance method for ovarian cancer. RRSO has been shown to not only reduce the risk of developing ovarian cancer but also to reduce cancer mortality. It is known that a certain percentage of occult cancer that could not be diagnosed preoperatively, an early cancer lesion called serous tubal intraepithelial carcinoma (STIC), or precancerous lesions called serous tubal intraepithelial lesion (STIL) and p53 signature are seen in fallopian tubes and ovaries. However, the clinical significance of STIC, STIL, and p53 signature is not well established. In this chapter, we summarize the reports about these lesions found in RRSO specimens with the aim of helping clinical management and future research.

Keywords Risk-reducing salpingo-oophorectomy (RRSO) · Sectioning and extensively examining the fimbriated end (SEE-FIM) protocol · High-grade serous carcinoma (HGSC) · Occult cancer · Serous tubal intraepithelial carcinoma (STIC) · Serous tubal intraepithelial lesion (STIL) · p53 signature · *TP53*

1 Introduction

To reduce the risk of ovarian cancer, risk-reducing salpingo-oophorectomy (RRSO) is recommended for women with a known *BRCA1/2* pathogenic variant, typically between 35 and 40 years, and upon completion of childbearing [1]. Since 2020 in

K. Masuda (✉)
Department of Obstetrics and Gynecology, Keio University School of Medicine, Tokyo, Japan
e-mail: ma-su-ken.a2@keio.jp

D. Aoki
Akasaka Sanno Medical Center, International University of Health and Welfare Graduate School, Minato-ku, Tokyo, Japan

D. Aoki et al. (eds.), *Practical Guide to Hereditary Breast and Ovarian Cancer*,
https://doi.org/10.1007/978-981-99-5231-1_4

Japan, women diagnosed as hereditary breast and ovarian cancer (HBOC) and who developed breast cancer are able to undergo RRSO as part of national health insurance. Therefore, it is expected that RRSO will be performed more frequently in Japan. A guide for performing RRSO by national health insurance is summarized on the webpage of the Japan Society of Gynecologic Oncology [2].

When the fallopian tubes and ovaries removed by RRSO are properly evaluated, an occult cancer that could not be diagnosed preoperatively or an early cancer lesion called serous tubal intraepithelial carcinoma (STIC) may be found. In addition, precancer lesions, such as serous tubal intraepithelial lesion (STIL) or p53 signature, may be found. However, there is no consensus on the clinical significance and management of these early lesions. In this article, we briefly described HBOC, ovarian cancer, and RRSO, then summarized the pathological evaluation of RRSO specimens, and what is known about STIC and precancerous lesions found in RRSO specimens.

2 HBOC and Ovarian Cancer

Women with pathological variants of the *BRCA1/2* gene have been shown to have an increased risk of developing cancers of the ovary, fallopian tube, and peritoneum. A meta-analysis showed the mean cumulative ovarian cancer risks of women with *BRCA1/2* variant at age 70 years were 40% (95% CI = 35–46%) for *BRCA1* and 18% (95% CI = 13–23%) for *BRCA2* variant carriers [3]. A large prospective cohort study of 6036 *BRCA1* and 3820 *BRCA2* female variant carriers showed that the cumulative ovarian cancer risks to age 80 years were 44% (95% CI = 36–53%) for *BRCA1* and 17% (95% CI = 11–25%) for *BRCA2* carriers [4]. In Japan, a large case-control study of 63,828 patients with 14 common cancer types and 37,086 controls from multi-institutional hospital-based registry showed the cumulative risk of ovarian cancer to age 85 years was estimated 65.6% (95% CI = 12.8–86.4%) for carriers of pathogenic variants in *BRCA1* and 14.8% (95% CI = 4.6–23.9%) for carriers of pathogenic variants in *BRCA2* [5].

Of four main histologic subtypes of epithelial ovarian cancer, high-grade serous carcinoma (HGSC) had the highest prevalence in women with pathological variants of *BRCA1/2*. HGSC is the most common and most lethal ovarian malignancy. HGSC shows papillary or solid growth patterns in histology, and most HGSC cases have *TP53* mutation as a key molecular abnormality [6]. As there are no effective screening methods and symptoms with ovarian cancer are minimal, most patients with HGSC are diagnosed as advanced disease and have a poor prognosis [7, 8].

HGSC is thought to arise from the distal fallopian tube, whereas other subtypes of ovarian cancer are thought to arise from ovarian surface epithelium or cortical inclusion cysts. Serous tubal intraepithelial carcinomas (STIC) found in the distal fallopian tube are suspected to be the precursor lesion of HGSC with sharing *TP53* mutations as well as p53 abnormal immunohistochemistry(IHC) in STICs and HGSC [9–11].

3 Significance of RRSO and Pathological Examination

The lack of effective methods for early detection of ovarian cancer and the poor prognosis in advanced ovarian cancer have led to the recommendation of RRSO for women with pathological variants of the *BRCA1/2*. The effectiveness of RRSO in reducing the risk of ovarian cancer in women with pathological variants of the *BRCA1/2* has been reported in a meta-analysis showing an approximately 80% reduction (HR = 0.21; 95% CI = 0.12–0.39) in the risk of ovarian or fallopian tube cancer by RRSO [12]. An observational study of 5783 women with *BRCA1/2* mutation showed that RRSO reduced not only risk of ovarian, fallopian tube, and peritoneal cancer by 80% (HR = 0.20; 95% CI = 0.13–0.30), but also all-cause mortality by 77% (HR = 0.23; 95% CI = 0.13–0.39) [13]. The NCCN guideline recommend that RRSO is performed typically between 35 and 40 years for women with *BRCA1* pathogenic variant. Since ovarian cancer onset tends to be later in women with *BRCA2* pathogenic variant, it is reasonable to delay RRSO between 40 and 45 years of age in women with *BRCA2* pathogenic variant, unless age at diagnosis of ovarian cancer in the family is earlier age [1].

In studies of women with *BRCA1/2* pathogenic variant who underwent RRSO, it has been reported that occult cancers and precancerous lesions called STIC were identified in fallopian tubes and ovaries removed by RRSO and in some cases only detected by pathologic examination of specimens. If occult cancer is identified in the RRSO sample, the patient requires further clinical management as a cancer patient such as a surgical staging procedure, thus RRSO specimens require particularly careful pathological evaluation. "Sectioning and Extensively Examining the FIMbriated end" (SEE-FIM) protocol have been developed to maximize the detection of small occult cancers and precancerous lesions for pathologic assessment of fallopian tubes. The protocol is as follows: The fimbriated end is sectioned longitudinally and combined with the remainder of the tube sectioned at 2- to 3-mm intervals [14, 15]. The ovaries should also be carefully sectioned, processed, and assessed [16].

4 Occult Cancer in RRSO Specimens

Several studies showed occult cancers that could not be diagnosed preoperatively might be found in the fallopian tubes and ovaries removed by RRSO. The incidence of occult cancers varies between 2 and 12% and is influenced by the age of the patients at surgery, the type and quality of screening before RRSO, the adequacy of the surgery, and the histopathological examination [13, 16–22]. Regarding the age of the patients at surgery, a report of 5783 women in *BRCA1/BRCA2* who underwent RRSO showed the prevalence of ovarian, fallopian tube, and peritoneal cancers found during RRSO was 1.5% for those younger than 40 years of age, 3.8% in those between the ages of 40 and 49 years, and 5–7% in those between 50 and 60 years, which increased with age [13]. A meta-analysis including thirty-four articles showed 61.3% of occult cancers occurred in the fallopian tubes and 32.3% in the ovaries, and 81.5% were in the early stages [23]. In Japan, the frequency of occult cancers found during RRSO was reported 2.6% and 10%, which is consistent with previous data [24, 25].

When occult cancer is identified, the patient requires further clinical management as a cancer patient; therefore, the risk of occult cancer identification should be explained for the patient in advance when performing RRSO.

5 Precancerous Lesions in RRSO Specimens

5.1 STIC

STIC lesions were originally identified in the fallopian tubes of patients undergoing RRSO for women with *BRCA1/2* pathogenic variant [26, 27]. The diagnostic process for STIC requires the identification of abnormalities in the fallopian tube epithelium. A systematic review article showed the six most frequently mentioned criteria for abnormal morphological features of STIC were (1) loss of polarity, (2) nuclear pleomorphism/atypia, (3) high nuclear-to-cytoplasmic ratio, (4) mitotic activity, (5) pseudostratification, and (6) prominent nucleoli [28]. In WHO classification, STIC is characterized by abnormal morphological features (high N/C ratio, nuclear enlargement, pleomorphism, hyperchromasia, lack of ciliated cells, loss of polarity with or without epithelial stratification, and occasional mitotic figures), aberrant p53 expression, and increased Ki-67 immunostaining (>10%) [29].

In RRSO specimens, the incidence of STIC varies from 0% to 11.5% (Table 1), which wide range could be related to the small sample size of the published studies, the variability in definition of STIC, the varying age of patients and percentage of *BRCA1/2* pathogenic variant, and the rare event of STIC [30–36].

5.1.1 STIC in RRSO Specimens

There is currently no standardized management for STICs identified in RRSO specimens. In NCCN guideline, management options consist of (1) observation alone with or without CA-125 testing when no evidence of invasive cancer is noted and (2) surgical staging with observation or chemotherapy if invasive cancer is noted. It is not clear whether surgical staging and/or adjuvant chemotherapy is beneficial for women with STIC [37]. ESMO guideline mentioned peritoneal restaging should be considered in cases of incidentally detected, apparently isolated STIC lesions, although the level of evidence is low [38].

It has been suggested that the risk of primary peritoneal cancer may be increased when STIC is found in RRSO specimens. In a systematic review of 78 cases of STIC in RRSO specimens, 3 cases (4.5%) were found to have primary peritoneal

Table 1 Number of patients and percentages with STIC, STIL, and p53 signature and criteria for pathological evaluation

		STIC	STIL	p53 signature	Criteria		
Reference	Subjects (*BRCA* status)	*N* of patients with STIC/*N* of total patients (% of STIC)	*N* of patients with STIL/*N* of total patients (% of STIL)	*N* of patients with p53 signature/*N* of total patients (% of p53 signature)	Atypia	p53 positive	Ki67 positive
Shaw et al. (2009) [11]	$n = 176$ (103 *BRCA1* and 73 *BRCA2*)	15/176 (8%)	3/176 (1.7%)	19/176 (10.8%)	A positive score for atypia required at least moderate mucosal epithelial proliferation, defined as including cellular crowding, stratification, loss of nuclear polarity, and at least moderate nuclear atypia in a discrete focus	p53 positive if greater than 75% of nuclei were stained positively in a region exceeding 12 cells	A positive score of Ki67 was assigned if there was nuclear staining in a discrete focus greater than twice that of adjacent epithelium
Leonhardt et al. (2011) [30]	$n = 14$ (with *BRCA1/2* mutations)	0/14 (0%)	1/14 (7.1%)	5/14 (35.7%)	Significant atypia, architectural alterations	p53 accumulation in 12 or more consecutive p53-positive secretory cell nuclei	A positive score for MIB1 required intense nuclear staining in over 80% of the nuclei
Wethington et al. (2013) [32]	$n = 593$ (189 *BRCA1* and 186 *BRCA2*)	12/593 (2%)	NR	NR	Nuclear/cytoplasmic ratio, nuclear pleomorphism, epithelial stratification with loss of polarity, irregular epithelial thickness, and exfoliation of cells into the tubal lumen	Null phenotype or >60% nuclear cell staining	Elevated Mib-1 (>15% nuclear cell staining)

(continued)

Table 1 (continued)

Reference	Subjects (*BRCA* status)	STIC *N* of patients with STIC/*N* of total patients (% of STIC)	STIL *N* of patients with STIL/*N* of total patients (% of STIL)	p53 signature *N* of patients with p53 signature/*N* of total patients (% of p53 signature)	Criteria Atypia	p53 positive	Ki67 positive
Cass et al. (2014) [33]	*n* = 78 (52 *BRCA1* and 26 *BRCA2*)	9/78 (11.5%)	15/78 (19.2%)	23/78 (29.5%)	Epithelial atypia/low-grade dysplasia was defined as mild nucleomegaly with inconspicuous nucleoli, mildly increased nuclear-to-cytoplasmic ratio, and/or slight loss of cellular polarity	At least six consecutive p53-positive tubal epithelial cells	Ki67 staining in >50% of nuclei
Ricciardi et al. (2017) [34]	*n* = 411 (157 *BRCA1*, 119 *BRCA2*, 14 *BRCA1/2*, and 60 *BRCA*-negative)	7/411 (1.7%)	16/411 (3.9%)	NR	NR	NR	NR
Stanciu et al. (2019) [35]	*n* = 300 (124 *BRCA1*, 118 *BRCA2*, 2 both *BRCA1/2*, and 29 no *BRCA* mutation)	7/300 (2.3%)	2/300 (0.7%)	75/300 (25%)	NR	NR	NR

Reference	Subjects (*BRCA* status)	STIC	STIL	p53 signature	Criteria		
		N of patients with STIC/*N* of total patients (% of STIC)	*N* of patients with STIL/*N* of total patients (% of STIL)	*N* of patients with p53 signature/*N* of total patients (% of p53 signature)	Atypia	p53 positive	Ki67 positive
Saccardi et al. (2021) [36]	*n* = 153 (80 *BRCA1* and 73 *BRCA2*)	4/153 (2.6%)	6/153 (3.9%)	NR	NR	NR	NR

NR not reported

cancer. However, due to the lack of a large number of cases, its clinical significance is not yet clear [31, 39].

5.1.2 STIC Coexisting with HGSC

STIC has been seen not only in RRSO specimens but also in the fallopian tubes of patients with HGSC. Several papers have shown that when STIC and HGSC coexist, they share a common *TP53* mutation, and STIC is now considered to be an early lesion of HGSC [40, 41]. However, not all patients with HGSC have STIC lesions in the fallopian tubes. A meta-analysis of 10 studies with 1643 patients with HGSC reported concurrent STICs with HGSCs in 31% (95% CI, 17–46%) of the resected fallopian tubes with frequencies ranging from 11% to 61% across studies [42]. In papers examining whether there is a difference in the characteristics of HGSC with or without STIC, there was no difference in the prognosis or genomic profiling including somatic copy number aberrations, messenger RNA, and micro-RNA, of HGSC with or without STIC [43, 44].

One model to explain the carcinogenesis mechanism of HGSCs without STIC is "precursor escape" model. It is proposed that early non-malignant precursor cells from the fallopian tube harbors a *TP53* mutation are shed in the peritoneal cavity and become malignant intraperitoneally and transform into HGSCs (ovarian cancer, fallopian tube cancer, and primary peritoneal cancer) [45–47].

5.1.3 STIC in Benign Gynecological Specimens

STICs have been reported to be identified not only in patients undergoing RRSO with *BRCA1/2* pathogenic variant or patients with HGSC but also in fallopian tube specimens removed for benign gynecological diseases. A population-based data in Canada showed STIC was detected in 8 (<0.1%) of 9392 cases with benign diagnoses [48]. In NCCN guideline, discovery of an STIC should prompt a genetics evaluation for women without prior genetic counseling and/or testing [37].

5.2 *STIL, p53 Signature*

Other aberrant lesions in fallopian tube epithelium have also been identified such as serous tubal intraepithelial lesions (STIL) and p53 signatures. Morphologically normal areas with aberrant p53 expression have been termed p53 signatures. STILs are areas of atypia that fall short of the criteria required for a diagnosis of STIC.

To distinguish STIC, STIL, and p53 signatures, a model has been developed that a combination of morphological suspicion of STIC and the results of p53 and Ki-67 stains [49]. Another approach presented is a decision tree that begins by identifying altered epithelium (SCOUTS), followed in subsequent steps by the presence of cilia, p53 immunostaining patterns, cell polarity, and finally atypia [47].

5.2.1 Frequencies of STIL and p53 Signature (Table 1)

More frequently seen in RRSO specimens is p53 signatures than STIL. The frequency of the p53 signature in RRSO samples was within the range of 11%–71% in previous reports [50–53]. A review article reported that an estimated incidence of p53 signatures was 16.2% (95% CI = 2.2–39.7%) and an estimated incidence of STIL was 1.6% (95% CI = 0.3–3.8%) in women with *BRCA1/2* pathogenic variant [28].

The p53 signature has been reported to be often found in fallopian tubes other than those of patients with *BRCA1/2* pathogenic variant. Of the 113 patients with benign gynecological/obstetric patients at low risk of ovarian cancer, the p53 signature was identified in 21 patients (19%). In this report, the incidence of the p53 signature was significantly lower in parous women and pregnant women [54]. Other reports showed no significant difference was observed in the frequency of p53 signature between BRCA mutation carriers and controls [33, 55].

5.2.2 Significance of p53 Signature and STIL

The clinical significance of the identification of the p53 signature and STIL in RRSO specimens is currently uncertain. To assess the pathological significance of the p53 signature, we conducted DNA sequencing for *TP53* variants of p53 signatures in 13 patients with pathogenic variants of *BRCA1/2* who underwent RRSO and in 17 control patients with the benign gynecologic disease. The proportions of pathogenic variants were significantly different between RRSO samples and controls ($p < 0.001$). These results suggest that the characteristics and risk of carcinogenesis might be different between p53 signatures in RRSO specimens and p53 signatures in controls. The result suggests that there might be 2 types of p53 signatures, one with a low risk of progression to STIC as seen in the control group and the other p53 signature with *TP53* pathological variants found in women with *BRCA1/2* pathogenic variants [55].

6 Future Prospective

Currently, RRSO is recommended for women with *BRCA1/2* pathogenic variants to reduce the risk of ovarian cancer after completion of childbearing in these women; however, it has become increasingly difficult to perform RRSO at the recommended age due to the recent aging of the childbearing years. Therefore, there is a need to develop other options.

One approach is to improve surveillance methods before RRSO. Since women harboring STIC or precancerous lesions with *TP53* pathogenic variant with *BRCA1/2* pathogenic variants are at high risk of developing ovarian cancer in the future, detection of STIC or p53 signature with *TP53* pathogenic variant may be applicable to early detection of ovarian cancer. In some studies, direct sampling of normal fallopian tubes and detection of precancerous and early invasive tumors by biomarkers detected in Pap smears and vaginal secretions have already been attempted [56, 57].

Another approach is biliteral salpingectomy with delayed oophorectomy as an alternative to RRSO, otherwise known as prophylactic salpingectomy with delayed oophorectomy (PSDO), since the fallopian tubes have been shown to be a major origin of ovarian cancer [58]. The risk of surgical menopause is expected to be avoided, while maintaining the risk reduction effect of ovarian cancer by removing fallopian tubes. The concern for the procedure is that women are still at risk of developing ovarian cancer. In addition, in premenopausal women, oophorectomy reduces the risk of developing breast cancer, but the magnitude is uncertain. To clarify these concerns, several large clinical trials are ongoing (NCT01907789, NCT02321228, and NCT04251052).

References

1. National Comprehensive Cancer Network NCCN Guidelines Version 1.2020: Genetic/Familial High-Risk Assessment: Breast, Ovarian, and Pancreatic. In: 2022. https://www.nccn.org/home/member. Accessed 25 Jun 2022.
2. The Japan Society of Gynecologic Oncology. https://jsgo.or.jp/opinion/05.html. Accessed 25 Jun 2022.
3. Chen S, Parmigiani G. Meta-analysis of BRCA1 and BRCA2 penetrance. J Clin Oncol. 2007;25:1329–33.
4. Kuchenbaecker KB, Hopper JL, Barnes DR, et al. Risks of breast, ovarian, and contralateral breast cancer for BRCA1 and BRCA2 mutation carriers. JAMA. 2017;317:2402–16.
5. Momozawa Y, Sasai R, Usui Y, et al. Expansion of cancer risk profile for BRCA1 and BRCA2 pathogenic variants. JAMA Oncol. 2022;8:871.
6. Lheureux S, Gourley C, Vergote I, Oza AM. Epithelial ovarian cancer. Lancet. 2019;393:1240–53.
7. Gilbert L, Basso O, Sampalis J, et al. Assessment of symptomatic women for early diagnosis of ovarian cancer: results from the prospective DOvE pilot project. Lancet Oncol. 2012;13:285–91.
8. Lim AWW, Mesher D, Gentry-Maharaj A, Balogun N, Jacobs I, Menon U, Sasieni P. Predictive value of symptoms for ovarian cancer: comparison of symptoms reported by questionnaire, interview, and general practitioner notes. J Natl Cancer Inst. 2012;104:114–24.
9. Callahan MJ, Crum CP, Medeiros F, Kindelberger DW, Elvin JA, Garber JE, Feltmate CM, Berkowitz RS, Muto MG. Primary fallopian tube malignancies in BRCA-positive women undergoing surgery for ovarian cancer risk reduction. J Clin Oncol. 2007;25:3985–90.
10. Powell BC, Kenley E, Chen LM, Crawford B, McLennan J, Zaloudek C, Komaromy M, Beattie M, Ziegler J. Risk-reducing salpingo-oophorectomy in BRCA mutation carriers: role of serial sectioning in the detection of occult malignancy. J Clin Oncol. 2005;23:127–32.

11. Shaw PA, Rouzbahman M, Pizer ES, Pintilie M, Begley H. Candidate serous cancer precursors in fallopian tube epithelium of BRCA1/2 mutation carriers. Mod Pathol. 2009;22:1133–8.
12. Rebbeck TR, Kauff ND, Domchek SM. Meta-analysis of risk reduction estimates associated with risk-reducing salpingo-oophorectomy in BRCA1 or BRCA2 mutation carriers. J Natl Cancer Inst. 2009;101:80–7.
13. Finch APM, Lubinski J, Møller P, et al. Impact of oophorectomy on cancer incidence and mortality in women with a BRCA1 or BRCA2 mutation. J Clin Oncol. 2014;32:1547–53.
14. Medeiros F, Muto MG, Lee Y, Elvin JA, Callahan MJ, Feltmate C, Garber JE, Cramer DW, Crum CP. The tubal fimbria is a preferred site for early adenocarcinoma in women with familial ovarian cancer syndrome. Am J Surg Pathol. 2006;30:230–6.
15. Crum CP, Drapkin R, Miron A, Ince TA, Muto M, Kindelberger DW, Lee Y. The distal fallopian tube: a new model for pelvic serous carcinogenesis. Curr Opin Obstet Gynecol. 2007;19:3–9.
16. Kauff ND, Barakat RR. Risk-reducing salpingo-oophorectomy in patients with germline mutations in BRCA1 or BRCA2. J Clin Oncol. 2007;25:2921–7.
17. Lamb JD, Garcia RL, Goff BA, Paley PJ, Swisher EM. Predictors of occult neoplasia in women undergoing risk-reducing salpingo-oophorectomy. Am J Obstet Gynecol. 2006;194:1702–9.
18. Domchek SM, Friebel TM, Garber JE, et al. Occult ovarian cancers identified at risk-reducing salpingo-oophorectomy in a prospective cohort of BRCA1/2 mutation carriers. Breast Cancer Res Treat. 2010;124:195–203.
19. Powell CB, Chen LM, McLennan J, Crawford B, Zaloudek C, Rabban JT, Moore DH, Ziegler J. Risk-reducing salpingo-oophorectomy (RRSO) in BRCA mutation carriers experience with a consecutive series of 111 patients using a standardized surgical-pathological protocol. Int J Gynecol Cancer. 2011;21:846–51.
20. Manchanda R, Abdelraheim A, Johnson M, et al. Outcome of risk-reducing salpingo-oophorectomy in BRCA carriers and women of unknown mutation status. BJOG An Int J Obstet Gynaecol. 2011;118:814–24.
21. Sherman ME, Piedmonte M, Mai PL, et al. Pathologic findings at risk-reducing salpingo-oophorectomy: primary results from gynecologic oncology group trial GOG-0199. J Clin Oncol. 2014;32:3275–83.
22. Mingels MJJM, Roelofsen T, Van Der Laak JAWM, De Hullu JA, Van Ham MAPC, Massuger LFAG, Bulten J, Bol M. Tubal epithelial lesions in salpingo-oophorectomy specimens of BRCA-mutation carriers and controls. Gynecol Oncol. 2012;127:88–93.
23. Cheng A, Li L, Wu M, Lang J. Pathological findings following risk-reducing salpingo-oophorectomy in BRCA mutation carriers: a systematic review and meta-analysis. Eur J Surg Oncol. 2020;46:139–47.
24. Nomura H, Ikki A, Fusegi A, et al. Clinical and pathological outcomes of risk-reducing salpingo-oophorectomy for Japanese women with hereditary breast and ovarian cancer. Int J Clin Oncol. 2021;26:2331–7.
25. Kobayashi Y, Hirasawa A, Chiyoda T, et al. Retrospective evaluation of risk-reducing salpingo-oophorectomy for BRCA1/2 pathogenic variant carriers among a cohort study in a single institution. Jpn J Clin Oncol. 2021;51:213–7.
26. Piek JMJ, Van Diest PJ, Zweemer RP, et al. Dysplastic changes in prophylactically removed fallopian tubes of women predisposed to developing ovarian cancer. J Pathol. 2001;195:451–6.
27. Seidman JD. Serous tubal intraepithelial carcinoma localizes to the tubal-peritoneal junction: a pivotal clue to the site of origin of extrauterine high-grade serous carcinoma (ovarian cancer). Int J Gynecol Pathol. 2015;34:112–20.
28. Bogaerts JMA, Steenbeek MP, van Bommel MHD, Bulten J, van der Laak JAWM, de Hullu JA, Simons M. Recommendations for diagnosing STIC: a systematic review and meta-analysis. Virchows Arch. 2022;480:725–37.
29. WHO. WHO classification of tumours/female genital tumours. 5th ed. IARC; 2020.
30. Leonhardt K, Einenkel J, Sohr S, Engeland K, Horn LC. P53 signature and serous tubal in-situ carcinoma in cases of primary tubal and peritoneal carcinomas and serous borderline tumors of the ovary. Int J Gynecol Pathol. 2011;30:417–24.

31. Patrono MG, Iniesta MD, Malpica A, Lu KH, Fernandez RO, Salvo G, Ramirez PT. Clinical outcomes in patients with isolated serous tubal intraepithelial carcinoma (STIC): a comprehensive review. Gynecol Oncol. 2015;139:568–72.
32. Wethington SL, Park KJ, Soslow RA, et al. Clinical outcome of isolated serous tubal intraepithelial carcinomas (STIC). Int J Gynecol Cancer. 2013;23:1603–11.
33. Cass I, Walts AE, Barbuto D, Lester J, Karlan B. A cautious view of putative precursors of serous carcinomas in the fallopian tubes of BRCA mutation carriers. Gynecol Oncol. 2014;134:492–7.
34. Ricciardi E, Tomao F, Aletti G, et al. Risk-reducing salpingo-oophorectomy in women at higher risk of ovarian and breast cancer: a single institution prospective series. Anticancer Res. 2017;37:5241–8.
35. Stanciu PI, Ind TEJ, Barton DPJ, Butler JB, Vroobel KM, Attygalle AD, Nobbenhuis MAE. Development of peritoneal carcinoma in women diagnosed with serous tubal intraepithelial carcinoma (STIC) following risk-reducing Salpingo-oophorectomy (RRSO). J Ovarian Res. 2019;12:1–6.
36. Saccardi C, Zovato S, Spagnol G, et al. Efficacy of risk-reducing salpingo-oophorectomy in BRCA1–2 variants and clinical outcomes of follow-up in patients with isolated serous tubal intraepithelial carcinoma (STIC). Gynecol Oncol. 2021;163:364–70.
37. National Comprehensive Cancer Network. NCCN Guidelines Version 1.2022: Ovarian Cancer; 2022
38. Colombo N, Sessa C, Du Bois A, et al. ESMO-ESGO consensus conference recommendations on ovarian cancer: pathology and molecular biology, early and advanced stages, borderline tumours and recurrent disease. Int J Gynecol Cancer. 2019;29:728–60.
39. Harmsen MG, Piek JMJ, Bulten J, et al. Peritoneal carcinomatosis after risk-reducing surgery in BRCA1/2 mutation carriers. Cancer. 2018;124:952–9.
40. Kuhn E, Kurman RJ, Vang R, Sehdev AS, Han G, Soslow R, Wang TL, Shih IM. TP53 mutations in serous tubal intraepithelial carcinoma and concurrent pelvic high-grade serous carcinoma-evidence supporting the clonal relationship of the two lesions. J Pathol. 2012;226:421–6.
41. Shih IM, Wang Y, Wang TL. The origin of ovarian cancer species and precancerous landscape. Am J Pathol. 2021;191:26–39.
42. Chen F, Gaitskell K, Garcia MJ, Albukhari A, Tsaltas J, Ahmed AA. Serous tubal intraepithelial carcinomas associated with high-grade serous ovarian carcinomas: a systematic review. BJOG An Int J Obstet Gynaecol. 2017;124:872–8.
43. Ducie J, Dao F, Considine M, Olvera N, Shaw PA, Kurman RJ, Shih IM, Soslow RA, Cope L, Levine DA. Molecular analysis of high-grade serous ovarian carcinoma with and without associated serous tubal intra-epithelial carcinoma. Nat Commun. 2017;8:990.
44. Boerner T, Walch HS, Nguyen B, et al. Exploring the clinical significance of serous tubal intraepithelial carcinoma associated with advanced high-grade serous ovarian cancer: a memorial Sloan Kettering team ovary study. Gynecol Oncol. 2021;160:696–703.
45. Soong T, Kolin D, Teschan N, Crum C. Back to the future? The fallopian tube, precursor escape and a dualistic model of high-grade serous carcinogenesis. Cancers (Basel). 2018;10:468.
46. Soong TR, Howitt BE, Miron A, et al. Evidence for lineage continuity between early serous proliferations (ESPs) in the fallopian tube and disseminated high-grade serous carcinomas. J Pathol. 2018;246:344–51.
47. Meserve EEK, Brouwer J, Crum CP. Serous tubal intraepithelial neoplasia: the concept and its application. Mod Pathol. 2017;30:710–21.
48. Samimi G, Trabert B, Geczik AM, Duggan MA, Sherman ME. Population frequency of serous tubal intraepithelial carcinoma (STIC) in clinical practice using SEE-Fim protocol. JNCI Cancer Spectr. 2018;2 https://doi.org/10.1093/JNCICS/PKY061.
49. Vang R, Visvanathan K, Gross A, et al. Validation of an algorithm for the diagnosis of serous tubal intraepithelial carcinoma. Int J Gynecol Pathol. 2012;31:243–53.
50. Lee Y, Miron A, Drapkin R, et al. A candidate precursor to serous carcinoma that originates in the distal fallopian tube. J Pathol. 2007;211:26–35.

51. Folkins AK, Jarboe EA, Saleemuddin A, Lee Y, Callahan MJ, Drapkin R, Garber JE, Muto MG, Tworoger S, Crum CP. A candidate precursor to pelvic serous cancer (p53 signature) and its prevalence in ovaries and fallopian tubes from women with BRCA mutations. Gynecol Oncol. 2008;109:168–73.
52. Visvanathan K, Vang R, Shaw P, Gross A, Soslow R, Parkash V, Shih IM, Kurman RJ. Diagnosis of serous tubal intraepithelial carcinoma based on morphologic and immunohistochemical features: A reproducibility study. Am J Surg Pathol. 2011;35:1766–75.
53. Mehra KK, Chang MC, Folkins AK, Raho CJ, Lima JF, Yuan L, Mehrad M, Tworoger SS, Crum CP, Saleemuddin A. The impact of tissue block sampling on the detection of p53 signatures in fallopian tubes from women with BRCA 1 or 2 mutations (BRCA+) and controls. Mod Pathol. 2011;24:152–6.
54. Ida T, Fujiwara H, Kiriu T, Taniguchi Y, Kohyama A. Relationship between the precursors of high grade serous ovarian cancer and patient characteristics: decreased incidence of the p53 signature in pregnant women. J Gynecol Oncol. 2019;30:e96.
55. Akahane T, Masuda K, Hirasawa A, et al. TP53 variants in p53 signatures and the clonality of STICs in RRSO samples. J Gynecol Oncol. 2022;33:e50.
56. Kinde I, Bettegowda C, Wang Y, et al. Evaluation of DNA from the papanicolaou test to detect ovarian and endometrial cancers. Sci Transl Med. 2013;5:167ra4. https://doi.org/10.1126/scitranslmed.3004952.
57. Erickson BK, Kinde I, Dobbin ZC, et al. Detection of somatic TP53 mutations in tampons of patients with high-grade serous ovarian cancer. Obstet Gynecol. 2014;124:881–5.
58. Nebgen DR, Hurteau J, Holman LL, Bradford A, Munsell MF, Soletsky BR, Sun CC, Chisholm GB, Lu KH. Bilateral salpingectomy with delayed oophorectomy for ovarian cancer risk reduction: A pilot study in women with BRCA1/2 mutations. Gynecol Oncol. 2018;150:79–84.

Mutual Feedback Between Gynecological HBOC Practice and Cancer Genomic Medicine

Ayumi Shikama

Abstract This article discusses the practice of gynecological hereditary breast and ovarian cancer (HBOC) and cancer genomic medicine in Japan. Recently, two significant changes in the field were emerging, including the approval of poly (ADP-ribose) polymerase (PARP) inhibitors for ovarian cancer and the inclusion of HBOC practice in public insurance coverage. Especially, the clinical use of PARP inhibitors for patients with ovarian cancer has been increasing, given the evidence of prognostic improvements in previous clinical trials. According to the increase in PARPi treatment, the biomarkers that determine their efficacy, including platinum-sensitivity status, *BRCA1/2* testing, and SNP-based genomic scar assays, have been widely adopted for clinical testing. Unfortunately, there is no biomarker that accurately reflects the tumor behaviors when the treatment is needed. The development of a biomarker that can predict the effect of PARP inhibitors in real time must be desired.

Keywords PARP inhibitors · Homologous recombination deficiency (HRD) · Platinum-sensitive ovarian cancer · *BRCA1/2* testing · SNP-based genomic scar assays

1 Introduction

In recent years, there have been two major changes in gynecological HBOC practice. First, poly (ADP-ribose) polymerase (PARP) inhibitors were approved for ovarian cancer. Second, HBOC practice was partially covered by public insurance, including *BRCA* genetic testing for all patients with ovarian cancer. That increases

---

A. Shikama (✉)
Department of Obstetrics and Gynecology, Institute of Medicine, University of Tsukuba, Tsukuba, Ibaraki, Japan
e-mail: ashikama@md.tsukuba.ac.jp

D. Aoki et al. (eds.), *Practical Guide to Hereditary Breast and Ovarian Cancer*, https://doi.org/10.1007/978-981-99-5231-1_5

the opportunities for gynecologic oncologists in Japan to be involved with HBOC women in daily practice. This article outlines the gynecological HBOC practice and cancer genomic medicine.

2 Characteristics of HBOC Ovarian Cancer

Ovarian cancer has been increasing due to lifestyle changes. It is one of the cancers with a poor prognosis, with a 5-year survival rate of 60% [1]. The distribution of histopathologic types in ovarian cancer is different between Japan and abroad [2, 3]. Although high-grade serous cancers are the most common in Europe and the United States, high-grade serous cancers and clear cell cancers are equally common in Japan. Clear cell cancers and endometrioid cancers are arising from endometriosis and have a high incidence of somatic mutations in ARID1A and PIK3CA [4]. On the other hand, TP53 somatic mutations are detected in about 90% of high-grade serous cancers [5, 6]. In addition, somatic mutations in homologous recombination repair genes such as *BRCA1/2*, ATM, and PALB2 are also frequently detected [7]. Since serous cancers were incidentally found in the fallopian tube when an HBOC woman underwent risk-reducing salpingo-oophorectomy, the fallopian tube is now considered to be the origin of serous cancers [8].

The frequency of HBOC-related ovarian cancer in Japan is about 15% of all ovarian cancers (about 10% for *BRCA1* and about 5% for *BRCA2*), but it is about 30% of high-grade serous cancers and about 25% of advanced stage III-IV disease [9]. HBOC-related ovarian cancers do not differ from those of sporadic ovarian cancers, and clinicians cannot figure out candidates of HBOC patients; all ovarian cancer patients are recommended *BRCA1/2* gene testing for the purpose of HBOC diagnosis. In these days, the change that *BRCA* testing covered by public health insurance in Japan makes the ovarian cancer patients access easier to genetic testing.

3 PARP Inhibitor

In 2005, it was first reported that PARP inhibitors showed specific antitumor effects against *BRCA1/2* mutant cancer cells [10]. *BRCA1/2* is one of the homologous recombination repair genes which repair DNA double-strand break. PARP works in the base excision repair of DNA single-strand breaks. PARP inhibitors prevent single-stranded DNA breaks repair. In *BRCA1/2*-mutant cells, double-stranded DNA is also not repaired, and cellular death is induced. This drug mechanism is called synthetic lethality (Fig. 1). Another antitumor effect of PARP inhibitors is that PARP trapping, which forms a PARP-DNA complex, causes PARP to keep at the DNA break site, thereby displacing the site where DNA repair factors work [11]. Currently, several PARP inhibitors are in development, and it has been reported that there is a correlation between PARP trapping and the antitumor effect of each PAPR

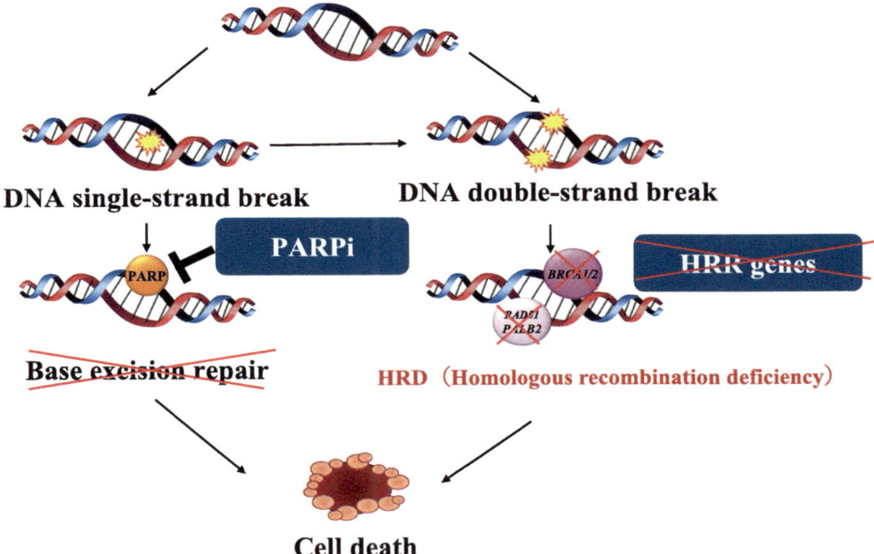

Fig. 1 Synthetic lethality

inhibitor. Furthermore, the allosteric effects of PARP inhibitors have been found to cause differences in the affinity of PARP-DNA binding, suggesting that they may cause differences in each PAPR inhibitor [12].

Considering the mechanism of PARP inhibitors, not only *BRCA1/2* but also homologous recombination repair genes such as ATM, PALB2, BRIP1, CHK1/2, and RAD51C/D can cause the failure of double-stranded DNA damage repair, and thus, PARP inhibitors are expected to be effective. Therefore, evaluating homologous recombination deficiency (HRD) may allow for an appropriate selection of patients to receive PARP inhibitors.

4 Biomarkers of PARP Inhibitors

4.1 Platinum-Sensitivity Status

HRD tumors have been shown to be highly sensitive to platinum-based chemotherapy, which is commonly used as a standard treatment for patients with ovarian cancer [13]. Platinum-sensitivity status has been established as a biomarker to predict PARP inhibitor benefit.

The Study19 evaluated the efficacy of olaparib versus placebo as a maintenance treatment in patients with platinum-sensitive, recurrent ovarian cancer in complete or partial response after platinum-based chemotherapy. Of 265 enrolled patients,

patients in the olaparib group had a significantly longer median duration of progression-free survival than those in the placebo group (8.4 vs. 4.8 months; HR 0.35; 95% confidence interval [CI], 0.25–0.49; $P < 0.001$) [14]. The benefit of PARP inhibitors as maintenance treatment in patients with platinum-sensitive, recurrent ovarian cancer was also reported for the use of niraparib in the NOVA study [15] and rucaparib in the ARIEL3 study [16].

4.2 BRCA1/2 *Testing*

In clinical trials investigating PARP inhibitor maintenance therapy in frontline and platinum-sensitive recurrent ovarian cancer, it is consistent that the subgroup of patients with *BRCA1/2* mutations had the greatest benefit from PARP inhibitor maintenance therapy. Within the SOLO1 study investigating the efficacy of olaparib maintenance therapy in frontline ovarian cancer patients with *BRCA1/2* pathogenic variants, the patients in the olaparib group had a significantly longer median duration of progression-free survival than those in the placebo group (56.0 vs. 13.8 months; HR 0.33; 95% CI, 0.25–0.43; $P < 0.001$) [17].

The benefit of PARP inhibitors as maintenance treatment in frontline settings was also reported for the use of niraparib in the PRIMA study [18] and rucaparib in the ATHENA MONO study [19]. It is relevant that *BRCA1/2* testing exhibits good clinical validity by identifying the subgroup of ovarian cancer patients who have a great benefit from PARP inhibitor maintenance therapy.

Of all ovarian cancer patients, germline mutations of *BRCA1/2* are found in 15%, and somatic mutations of *BRCA1/2* are found in 5–7% [20, 21]. Despite limited data, a similar PARP inhibitor benefit was observed for the patients with both germline and somatic *BRCA1/2* mutations. In daily practice, BRACAnalysis® (Myriad) is utilized for detecting germline *BRCA1/2* mutations, and FoundationOne is utilized for detecting somatic mutations.

4.3 SNP-Based Genomic Scar Assays

Genomic scars are considered an indirect measure of HRD because tumors with HRD represent genomic instability induced by DNA repair deficiency. Several genomic scar assays have been evaluated in prior clinical trials.

The most common SNP-based genomic scar assay is myChoice CDx® (Myriad), which combines tumor *BRCA* mutation and genomic instability (GI) score, which is calculated by large-scale transitions (LST), loss of heterozygosity (LOH), and telomeric allelic imbalance (TAI). GI score was scored on a scale of 0–100, and the cutoff score was set to be 42. Any tumors that scored $\geqq 42$ or had *BRCA1/2* mutations were considered to be HRD. The PRIMA study investigated the efficacy of niraparib maintenance therapy in frontline ovarian cancer patients and used myChoice CDx® (Myriad) as the biomarkers. Within the PRIMA study, niraparib maintenance therapy in the patients having a tumor with HRD is more effective than in those having a tumor with HRP [18]. Within the PAOLA study,

HRD determined by myChoice CDx® (Myriad) in the olaparib group had a significantly longer median duration of progression-free survival than those in the placebo group (56.0 vs. 13.8 months; HR 0.33; 95% CI, 0.25–0.43; $P < 0.001$) [20].

The other common SNP-based genomic scar assay is FoundationOne CDx® (FoundationOne Medicine), which evaluates the ratio of a genomic region with LOH determined through SNP sequencing. The cutoff has differed from 14% to 16% in each clinical trial.

The ATHENA study investigated whether rucaparib was effective as a first-line maintenance treatment for advanced ovarian cancer and used FoundationOne CDx® to evaluate the HRD status. The primary endpoint of investigator-assessed PFS was assessed in a step-down procedure first in the HRD population (the patients with *BRCA* mutant or LOH high carcinoma) and then in the intent-to-treat (ITT) population. Among the HRD population, the PFS was superior in the rucaparib group compared to the placebo group.

(28.7 vs. 11.3 months; HR 0.47; 95% CI, 0.31–0.72; $P = 0.0005$) [17]. In ATHENA mono, the effectiveness of rucaparib was also revealed in the ITT population (rucaparib vs. placebo; 20.2 vs. 9.2 months; HR 0.52; 95% CI, 0.40–0.68; $P < 0.0001$). However, the median PFS in the rucaparib group was longer in the HRD population than in the ITT population. Similar results were revealed in the ARIEL2 study [22] and the ARIEL3 study [16]. Rucaparib has not been approved in Japan.

Therefore, we only clinically use myChoice CDx® for companion diagnosis in primary advanced ovarian cancer.

5 New Candidates for Evaluating HRD Status

5.1 Homologous Recombination Repair (HRR)-Related Gene Mutations

Loss of function mutations in HRR-related genes, which encode important roles in the HRR pathway, including ATM, PALB2, BRIP1, CHK1/2, RAD51C/D, and Fanconi anemia genes, have been recognized as key causes of HRD in ovarian cancers [23]. ARIEL2 trials have shown that RAD51C mutations or methylations were associated with long-term responses to PARP inhibitor therapy [24]. Study 19 reported similar findings that HRR-related genes, including CDK12, RAD 51B, and BRIP 1 [25]. On the other hand, Takaya reported conflicting data that ATM, ATR, FANCA, CANCD2, FANCM, or PALB2 mutations were not associated with HRD or platinum sensitivity based on the analyses of ovarian cancer TCGA data [26]. So far, no HRR-related genes have been established as biomarkers for PARP inhibitor.

An increased lifetime risk of breast and ovarian cancer has been also reported for other HRD-related genes besides *BRCA1/2*. The identification of new HRD-related genes in the search for biomarkers for PARP inhibitors may lead to the identification of new disease susceptibility genes, morbidity risks, and preventive measures.

5.2 *Mutation Signatures*

The different mutation processes produce unique combinations of mutation types, called "mutation signatures." The profile of each signature is displayed using 6 substitution subtypes: C > A, C > G, C > T, T > A, T > C, and T > G. In addition, 96 possible mutations are considered by incorporating information on the immediate 5′ and 3′ bases of each mutated base. Mutation signatures are displayed and reported based on the observed 3-base frequency of the human genome. Each signature provides additional information such as the type of cancer for which the signature was discovered, the etiology of the mutational process underlying the signature, and characteristics of other mutations associated with each signature. Among 30 mutational signatures in COSMIC, Signature 3 (Fig. 2) has been reported to be associated with HRD [27]. Especially, Signature 3 is strongly associated with germline and somatic *BRCA1/2* mutations in breast, pancreatic, and ovarian cancers. The signature profile could be used as a biomarker for PARP inhibitor in the future.

Unfortunately, there is no biomarker that accurately reflects the tumor behaviors when the treatment is needed. The development of a biomarker that can predict the effect of PARP inhibitors in real time must be desired.

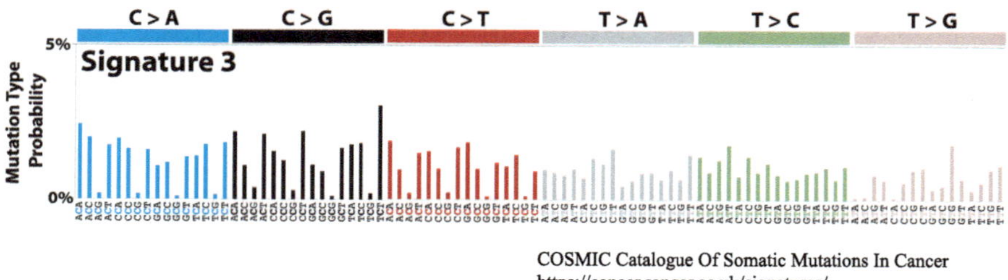

Fig. 2 Mutation signature 3

6 Clinical Use of PARP Inhibitors for Ovarian Cancer in Japan

Figure 3 shows clinical use of PARP inhibitor for ovarian cancer in Japan. The opportunities of PARP inhibitor treatment are increasing. Furthermore, several clinical trials for PARP inhibitor in ovarian cancer are ongoing, so it will be more widely used in the future.

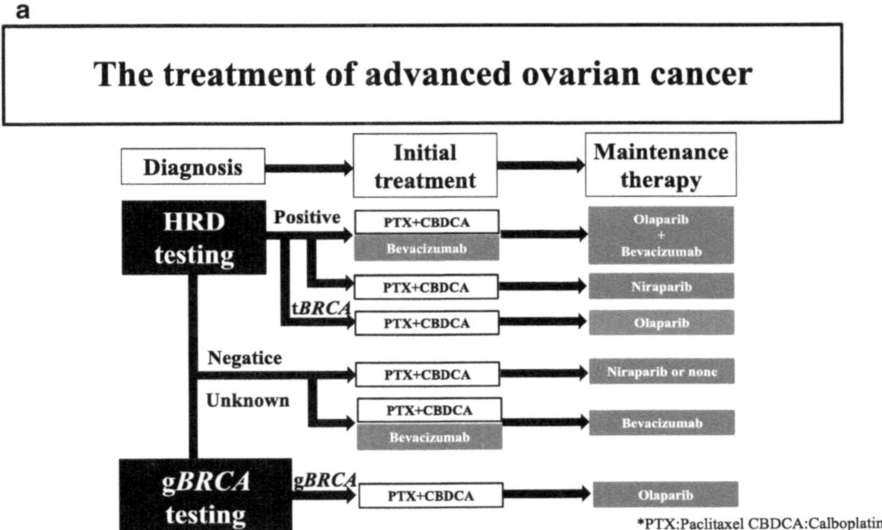

Fig. 3 (**a**) The treatment of advanced ovarian cancer (**b**) Treatment of recurrent ovarian cancer

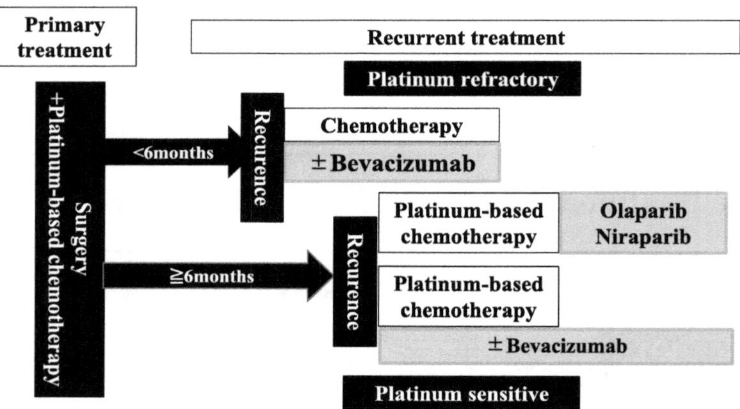

Fig. 3 (continued)

7 Conclusion

In recent years, ovarian cancer patients who were previously difficult to treat have come to benefit from PARP inhibitors. As in the development of PARP inhibitor, research of hereditary tumors is important not only to improve our understanding of the disease but also to provide the opportunity for novel therapies.

References

1. Https://Ganjoho.Jp/Reg_Stat/Statistics/Stat/Cancer/19_Ovary.Html. [Cited 15 Dec 2020].
2. Disaia P, Creasman W, Mannell R, Mcmeekin S, Mutch D. Clinical gynecologic oncology. 9th Edition. 2017 February 4, 2017.
3. Gynecology Tjsooa. Patient Annual Report for 2020. 2020.
4. Jones S, Wang TL, Shih Ie M, Mao TL, Nakayama K, Roden R, et al. Frequent mutations of chromatin remodeling gene Arid1a in ovarian clear cell carcinoma. Science. 2010;330(6001):228–31.
5. Cancer Genome Atlas Research Network. Integrated genomic analyses of ovarian carcinoma. Nature. 2011;474(7353):609–15.
6. Vang R, Levine DA, Soslow RA, Zaloudek C, Shih Ie M, Kurman RJ. Molecular alterations of Tp53 are a defining feature of ovarian high-grade serous carcinoma: a Rereview of cases lacking Tp53 mutations in the cancer genome atlas ovarian study. Int J Gynecol Pathol. 2016;35(1):48–55.
7. Konstantinopoulos PA, Ceccaldi R, Shapiro GI, D'andrea AD. Homologous recombination deficiency: exploiting the fundamental vulnerability of ovarian cancer. Cancer Discov. 2015;5(11):1137–54.

8. Kindelberger DW, Lee Y, Miron A, Hirsch MS, Feltmate C, Medeiros F, et al. Intraepithelial carcinoma of the fimbria and pelvic serous carcinoma: evidence for a causal relationship. Am J Surg Pathol. 2007;31(2):161–9.
9. Enomoto T, Aoki D, Hattori K, Jinushi M, Kigawa J, Takeshima N, et al. The first Japanese Nationwide multicenter study of Brca mutation testing in ovarian cancer: characterizing the cross-sectional approach to ovarian cancer genetic testing of Brca (Charlotte). Int J Gynecol Cancer. 2019;29(6):1043–9.
10. Farmer H, Mccabe N, Lord CJ, Tutt AN, Johnson DA, Richardson TB, et al. Targeting the Dna repair defect in Brca mutant cells as a therapeutic strategy. Nature. 2005;434(7035):917–21.
11. Murai J, Huang S-Y, Das BB, Renaud A, Zhang Y, Doroshow JH, et al. Trapping of Parp1 and Parp2 by clinical Parp inhibitors. Cancer Res. 2012;72(21):5588–99.
12. Zandarashvili L, Langelier MF, Velagapudi UK, Hancock MA, Steffen JD, Billur R, et al. Structural basis for allosteric Parp-1 retention on Dna breaks. Science. 2020;368(6486):eaax6367.
13. Pennington KP, Walsh T, Harrell MI, Lee MK, Pennil CC, Rendi MH, et al. Germline and somatic mutations in homologous recombination genes predict platinum response and survival in ovarian, fallopian tube, and peritoneal carcinomas. Clin Cancer Res. 2014;20(3):764–75.
14. Ledermann J, Harter P, Gourley C, Friedlander M, Vergote I, Rustin G, et al. Olaparib maintenance therapy in platinum-sensitive relapsed ovarian cancer. N Engl J Med. 2012;366(15):1382–92.
15. Mirza MR, Monk BJ, Herrstedt J, Oza AM, Mahner S, Redondo A, et al. Niraparib maintenance therapy in platinum-sensitive, recurrent ovarian cancer. N Engl J Med. 2016;375(22):2154–64.
16. Coleman RL, Oza AM, Lorusso D, Aghajanian C, Oaknin A, Dean A, et al. Rucaparib maintenance treatment for recurrent ovarian carcinoma after response to platinum therapy (Ariel3): a randomised, double-blind, placebo-controlled, phase 3 trial. Lancet. 2017;390(10106):1949–61.
17. Disilvestro P, Banerjee S, Colombo N, Scambia G, Kim B-G, Oaknin A, et al. Overall survival with maintenance Olaparib at a 7-year follow-up in patients with newly diagnosed advanced ovarian cancer and a Brca mutation: the Solo1/Gog 3004 trial. Journal of. Clin Oncol. 2022;78:25.
18. González-Martín A, Pothuri B, Vergote I, Depont Christensen R, Graybill W, Mirza MR, et al. Niraparib in patients with newly diagnosed advanced ovarian cancer. N Engl J Med. 2019;381(25):2391–402.
19. Monk BJ, Parkinson C, Lim MC, O'Malley DM, Oaknin A, Wilson MK, et al. Athena–mono (Gog-3020/Engot-Ov45): a randomized, double-blind, phase 3 trial evaluating rucaparib monotherapy versus placebo as maintenance treatment following response to first-line platinum-based chemotherapy in ovarian cancer. J Clin Oncol. 2022;40(17_Suppl):LBA5500.
20. Ray-Coquard I, Pautier P, Pignata S, Pérol D, González-Martín A, Berger R, et al. Olaparib plus bevacizumab as first-line maintenance in ovarian cancer. N Engl J Med. 2019;381(25):2416–28.
21. Callens C, Vaur D, Soubeyran I, Rouleau E, Just PA, Guillerm E, et al. Concordance between tumor and germline Brca status in high-grade ovarian carcinoma patients in the phase iii Paola-1/Engot-Ov25 trial. J Natl Cancer Inst. 2020;113:917.
22. Swisher EM, Lin K, Oza AM, Scott CL, Giordano H, Sun J, et al. Rucaparib in relapsed, platinum-sensitive high-grade ovarian carcinoma (Ariel2 part 1): An international, multicentre, open-label, phase 2 trial. Lancet Oncol. 2017;18(1):75–87.
23. Zhang H, Liu T, Zhang Z, Payne SH, Zhang B, Mcdermott JE, et al. Integrated Proteogenomic characterization of human high-grade serous ovarian cancer. Cell. 2016;166(3):755–65.
24. Swisher EM, Kwan TT, Oza AM, Tinker AV, Ray-Coquard I, Oaknin A, et al. Molecular and clinical determinants of response and resistance to Rucaparib for recurrent ovarian cancer treatment in Ariel2 (parts 1 and 2). Nat Commun. 2021;12(1):2487.
25. Hodgson DR, Dougherty BA, Lai Z, Fielding A, Grinsted L, Spencer S, et al. Candidate biomarkers of Parp inhibitor sensitivity in ovarian cancer beyond the Brca genes. British Journal of. Cancer. 2018;119(11):1401–9.

26. Takaya H, Nakai H, Takamatsu S, Mandai M, Matsumura N. Homologous recombination deficiency status-based classification of high-grade serous ovarian carcinoma. Sci Rep. 2020;10(1):2757.
27. Gulhan DC, Lee JJ, Melloni GE, Cortés-Ciriano I, Park PJ. Detecting the mutational signature of homologous recombination deficiency in clinical samples. Nat Genet. 2019;51(5):912–9.

Part III
Basic Science

Heterogeneities in Hereditary Cancer Genes as Revealed by a Large-Scale Genome Analysis

Yukihide Momozawa

Abstract *BRCA1* and *BRCA2* are considered top runners in personalized medicine. However, limited information on target cancer types, clinical interpretation of genetic variants, and prevalence in different populations prevents from taking advantage of its full potential in clinical practice. In this chapter, we reanalyzed all genetic data of >66,000 individuals on breast, prostate, pancreatic, colorectal, and renal cancers in the Japanese population to better compare these cancer types, with which we discussed about various heterogeneities to improve personalized medicine.

Keywords Hereditary cancer · *BRCA1* · *BRCA2* · Japanese · Heterogeneity · Targeted sequencing

1 Background

1.1 Limited Information on Personalized Medicine for BRCA1 *and* BRCA2

"Personalized medicine is an emerging practice of medicine that uses an individual's genetic profile to guide decisions made in regard to the prevention, diagnosis, and treatment of disease" according to the National Human Genome Research Institute in the United States. *BRCA1* and *BRCA2* are recognized as the top runners of personalized medicine. Genetic testing of these genes started in the 1990s [1], but not all necessary information, such as target cancer types, clinical interpretation of genetic variants, prevalence in different populations, and efficacy of poly (ADP-ribose) polymerase (PARP) inhibitors, was available.

Y. Momozawa (✉)
Laboratory for Genotyping Development, RIKEN Center for Integrative Medical Sciences (IMS), Yokohama City, Kanagawa, Japan
e-mail: momozawa@riken.jp

D. Aoki et al. (eds.), *Practical Guide to Hereditary Breast and Ovarian Cancer*,
https://doi.org/10.1007/978-981-99-5231-1_6

When *BRCA1* and *BRCA2* were identified, target cancer types were breast and ovarian cancers [2, 3]. Although prostate [4] and pancreatic cancers [5] are considered established, in addition to breast and ovarian cancers [6], more target cancer types have been proposed in various studies [7–12]. When considering the surveillance of carriers with pathogenic variants, target cancer types provide very important information. However, there is insufficient evidence to change clinical guidelines about *BRCA1/2*.

Interpretation of genetic variants is challenging. Due to next-generation sequencing, the patient's DNA is normally analyzed to sequence all coding regions of *BRCA1* and *BRCA2*, instead of genotyping some founder pathogenic variants shared by many patients. As a result, many genetic variants have been identified. However, doctors and patients want to know if there is a pathogenic variant that increases the risk of breast and ovarian cancers among the many genetic variants. As functional and in silico assays alone cannot determine pathogenic variants, classification systems using several types of data, such as its frequency in various populations, computational, functional, and segregation data, are used. The American College of Medical Genetics and Genomics and Association for Molecular Pathology (ACMG/AMP) guidelines are generally considered as standard [13], while the evidence-based network for the interpretation of germline mutant alleles (ENIGMA) guidelines are more specific for *BRCA1* and *BRCA2* [14]. However, a large proportion of genetic variants, particularly rare non-synonymous variants, could not be annotated. These are known variants of uncertain significance (VUS) [15]. VUS are considered a barrier to personalized medicine because a patient cannot have accurate information on genetic risk. To decrease the number of VUS, more information about clinical information and family history from patients and controls should be accumulated; this is an issue that is yet to be resolved. In addition, federated analysis was recently realized to prompt it by "bringing the code to the data": analyzing the sensitive patient-level data computationally within its secure home institution and providing researchers with valuable insights from data that would not otherwise be accessible [16].

The frequency of pathogenic variants in *BRCA1* and *BRCA2* differs across populations [17]. This difference is partially due to the presence of founder pathogenic variants. A founder pathogenic variant originating from one or more ancestors is observed with high frequency in a group that has been geographically or culturally isolated. For instance, the founder pathogenic variants 185delAG and 5382insC in *BRCA1* and 6174delT in *BRCA2* increased the frequency of pathogenic variants in *BRCA1* and *BRCA2* in Ashkenazi Jews [17]. Because of this high frequency, population screening has been proposed. The prevalence of pathogenic variants in various populations should therefore be determined to develop the best strategy to decrease the risk caused by both genes [18].

As discussed above, although genetic testing of *BRCA1/2* began more than a quarter of a century ago, the available information is still insufficient. This situation is even worse for other hereditary cancer genes as their information is limited for clinical use [19].

1.2 *Importance of Large-Scale Data*

Several concerns have been raised for personalized medicine with pathogenic variants in hereditary cancer genes. However, germline pathogenic variants could provide a risk estimation for each cancer type before the onset of each cancer and reveal a new mechanism to develop cancer, leading to the development of targeted drugs, such as PARP inhibitors [20]. Therefore, genetic analysis of rare pathogenic variants should be encouraged for various cancer types. In addition, the prevalence and clinical characteristics of each cancer type vary across populations and countries [21, 22], as well as the distribution and frequency of rare pathogenic variants [17]. The difficulty is that most pathogenic variants are very rare; the frequency of the most common pathogenic variant in *BRCA1* (p.Leu63*) is ~0.02% (=1/5000) in cancer-free controls [23]. Therefore, large-scale genetic data (i.e., several thousand samples or more) on the association between each cancer type and hereditary cancer genes should be investigated in each population/country. Two studies published in the New England Journal of Medicine in 2021 analyzed >100,000 samples regarding the association between breast cancer and 28–34 hereditary cancer genes [24, 25]. Although these associations have been investigated in several studies [19], the publication of the above-mentioned studies in this journal in 2021 analyzed >100,000 samples, which is considered clinically important by the New England Journal of Medicine.

2 Risk and Frequency of Pathogenic Variants in 27 Hereditary Cancer Genes

To overcome the difficulties as mentioned above and to improve personalized medicine, we performed targeted sequencing of 8–27 genes in 7051 cases with breast cancer [23], 7636 cases with prostate cancer [26], 1005 cases with pancreatic cancer [27], 12,503 cases with colorectal cancer [28], 1532 cases with renal cancer [29], and up to 23,705 controls in five different studies. All samples were collected from BioBank Japan between 2003 and 2018. In the present chapter, we reanalyzed all the data together, which enabled us to perform a direct comparison between the studies.

2.1 *BioBank Japan*

BioBank Japan was established as a multi-institutional hospital-based registry that collects DNA, plasma from peripheral blood leukocytes, and clinical information from patients with 51 common diseases [30–32], including 14 cancer types (biliary tract, breast, cervical, colorectal, endometrial, esophageal, gastric, liver, lung, lymphoma, ovarian, pancreatic, prostate, and kidney), from all over Japan. Most samples were mainly analyzed in a genome-wide association study (GWAS) [33], and the single-nucleotide polymorphism (SNP) array data and statistical summaries obtained were deposited at the National Bioscience Database Center (NBDC; https://humandbs.biosciencedbc.jp/hum0014-latest) and RIKEN (http://jenger.riken.jp/en/) databases. These data contribute to an international collaboration using meta-analysis to identify genome loci associated with various diseases and phenotypes [34].

2.2 Sequencing

The appearance of next-generation sequencing in 2006 allowed producing a much larger amount of genetic data with dramatically decreased costs. For whole-genome sequencing (WGS) of one individual, the cost decreased from $100 million in 2000 to $1000 in 2020 (https://www.genome.gov/about-genomics/fact-sheets/DNA-Sequencing-Costs-Data). In terms of time consumption, WGS took several years in 2000 but only 1–2 days in 2020. However, WGS for genetic testing is rare. Targeted sequencing is normally used to analyze specific genes. Although WGS provides sequence data from the entire genome, ~98% of the whole genome is not translated into protein sequences [35]. In addition, most of the genetic variants in these non-coding regions are nearly impossible to interpret using the current biological knowledge. The remaining 2% correspond to coding regions of approximately 20,000 genes but only a few of these genes are known to increase the risk of certain cancer types. Considering the time and cost of WGS, targeted sequencing to analyze the coding regions of specific genes is the most reasonable option for genetic testing.

Several targeted sequencing methods have been developed. Our group has used targeted sequencing based on multiplex PCR [36] (Fig. 1). First, we designed primer pairs to amplify specific targets. In our breast cancer study [23], for instance, we designed 471 primer pairs to cover all the coding regions of 11 genes not overlapping with known common variants. After optimization of the multiplex PCR with 300–1000 primer pairs in a single PCR reaction, we perform a multiplex PCR to amplify the target regions. As the second PCR, we added adaptor sequences, for use in the Illumina sequencer, and eight-base-pair (bp) barcoding sequences to distinguish each individual. A major advantage of this method is the use of a normal 384-well PCR plate and manipulation robot, which dramatically speeds up library preparation. We then pooled and sequenced all DNA libraries on the Illumina next-generation sequencers HiSeq2500 and NovaSeq6000. Based on the barcoding sequences, the obtained sequencing data were allocated to each individual. Bioinformatic analysis then revealed the genotyping data of each individual. Our custom scripts have been deposited at https://github.com/Laboratory-for-Genotyping-Development/TargetSequence.

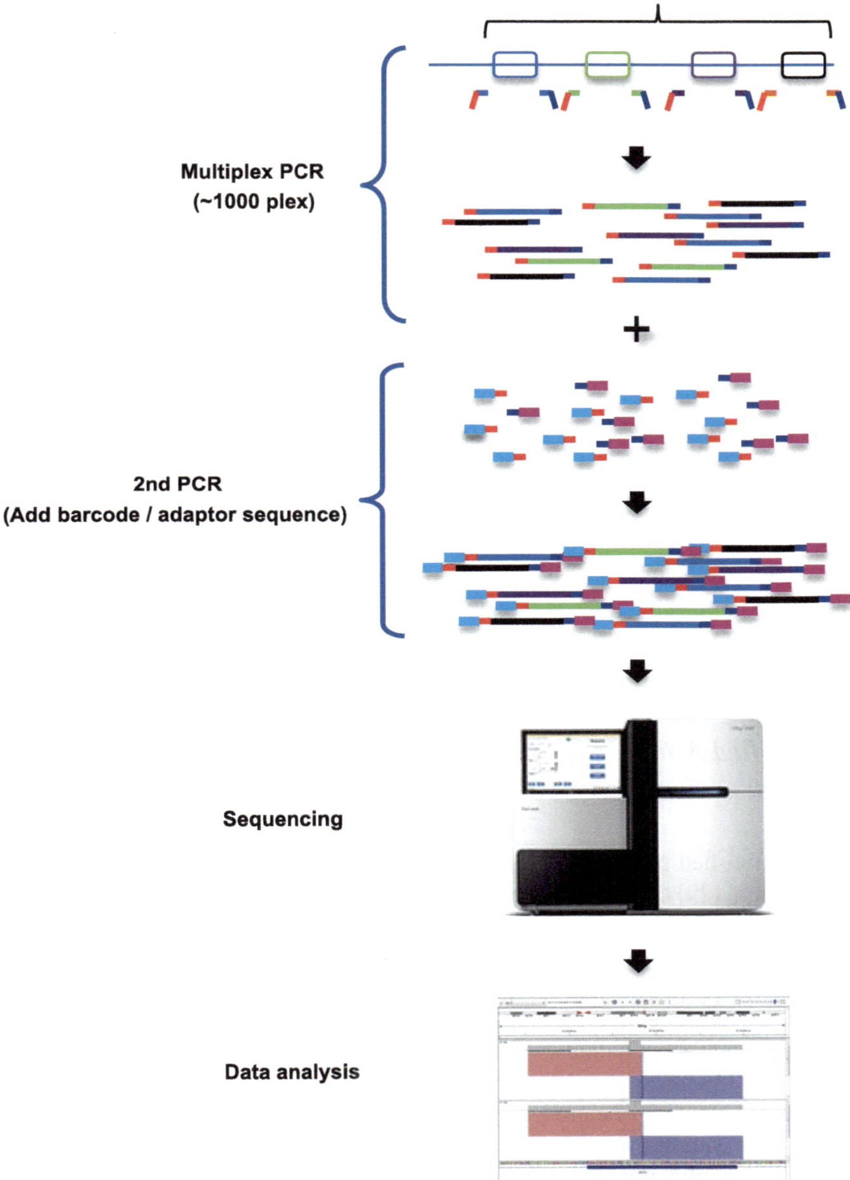

Fig. 1 Multiplex-PCR based targeted sequencing method our group developed. (1) We design primer pairs to cover all the cording regions of candidate genes, and we perform a multiplex PCR with 300–1000 primer pairs to amplify the target regions. (2) We add adaptor sequences for the Illumina sequencer and barcode sequences to distinguish each individual. (3) We pool and sequence all DNA libraries on the Illumina sequencer. (4) Based on the barcode sequences, each sequence read is allocated to each individual and bioinformatics analysis reveal genotyping data

2.3 Clinical Interpretation of Genetic Variants

Targeted sequencing identified genetic variants in the coding regions. However, genetic variants are largely divided into pathogenic variants, benign variants, and VUS based on various rules and guidelines. More detailed classifications have also been used, such as likely pathogenic, likely benign, protective, association, and drug response [37].

Guidelines established by the ACMG-AMP [13] include 16 evidence for pathogenic and 12 evidence for benign variants, consisting of association results, known clinical significance information, population data, computational data from in silico experiments, and functional data. However, because detailed methods are not defined, several inconsistencies between laboratories have been reported [38]. In addition, new information such as population frequency and functional test results and the revision of guidelines could change clinical significance [39], but there are very few examples of large changes between pathogenic and benign variants [40].

BRCA1/2-specific guidelines have also been developed using ENIGMA [14]. ClinVar collected interpretations of variants from researchers and genetic test companies [37]. As an alternative, loss of function variants such as nonsense and frameshift variants or pathogenic variants registered in ClinVar could be considered pathogenic. In this case, new pathogenic missense variants would be missed; however, this method is acceptable in a top clinical journal [24].

2.4 Unified Analysis of >66,000 Individuals with Five Cancer Types

We have published papers on breast [23], prostate [26], pancreatic [27], colorectal [28], and renal [29] cancers. However, several details of analytical methods, such as the number of genes, the number of controls, annotation of clinical significance, and association analysis, varied according to each study concept, design, and reviewers' requests, which hinders direct comparisons of cancer types. In the present study, we jointly reanalyzed 7091 female breast cancer patients, 7743 prostate cancer patients, 1009 pancreatic cancer patients, 12,578 colorectal cancer patients, 849 renal cancer patients, and 38,161 controls. To increase the statistical power, controls had no personal or family history of cancer. We considered the loss-of-function (LoF) variants or pathogenic variants deposited in ClinVar as pathogenic variants. Although we used data from all patients to calculate the carrier rate of pathogenic variants, only data from patients without a family history of cancer were used to estimate the disease risk more precisely as used in our recent study [41].

2.5 Risk and Frequency of Pathogenic Variants

Table 1 shows the carrier frequencies of the pathogenic variants in each gene for each cancer type. In female breast cancer patients, *BRCA2* (1.322%) and *BRCA1* (2.754%) showed a carrier frequency > 1%, as in other cancer types. For instance, *BRCA2* in prostate cancer (1.097%), *ATM* in pancreatic cancer (1.682%), and *BRCA2* in pancreatic cancer (2.374%).

Disease risk was also calculated again (Table 2). In female breast cancer, five genes showed $P < 0.05$: *BRCA2* [$P = 5.83 \times 10^{-24}$, odds ratio (OR) = 8.7, 95% confidence interval (CI): 5.7–13.3], *BRCA1* ($P = 4.98 \times 10^{-9}$, OR = 10.3, 95% CI: 4.7–22.4), *PTEN* ($P = 8.56 \times 10^{-4}$, OR = 15.5, 95% CI: 3.1–77.9), *CHEK2* ($P = 2.47 \times 10^{-3}$, OR = 3.5, 95% CI: 1.6–7.8), and *PALB2* ($P = 2.85 \times 10^{-3}$, OR = 4.0, 95% CI: 1.6–9.8). When these data were compared with the largest dataset of breast cancer (>113,000 women) [25], the results were similar; although OR was different, the 95% CI overlapped between both studies. In the original study [23], *TP53*, *NF1*, and *ATM* also showed $P < 0.05$. The differences could be explained by the lower statistical power in this current analysis, which focused only on cases without a family history. These three genes should be investigated in a larger sample size study to estimate disease risk.

In prostate cancer, we identified only *BRCA2* with $P < 0.05$ ($P = 8.06 \times 10^{-8}$, OR = 4.2, 95% CI: 2.5–7.1). In the original paper [26], *HOXB13* (OR = 4.73, 95% CI: 2.84–8.19) and *ATM* (OR = 2.86, 95% CI: 1.63–5.15) also showed $P < 0.05$. Note that *BRCA1* did not show $P < 0.05$, even in the original study with higher statistical power using controls without a family history of cancer. *BRCA1* and *BRCA2* are dealt with together very often, but there is a large difference in the risk of prostate cancer between *BRCA1* and *BRCA2*. This was observed in several populations [4, 42]. Currently, the Japanese insurance allows for the genetic testing of both genes, but the difference in therapeutic effects should be carefully investigated in future studies.

Regarding the inconsistency between the current calculations and that in the original study [26], missense variants in *HOXB13*, especially p.Gly132Glu, contributed to increased disease risk, and the current analysis failed to reveal this association because we focused only on the LoF and pathogenic variants in ClinVar. p.Gly132Glu was shared by individuals and was considered a founder pathogenic variant. Interestingly, different founder pathogenic variants have been observed: p.Gly84Glu in European [43] and p.Gly135Glu in Chinese [44] populations. A recent functional study of p.Gly84Glu revealed *HOXB13* recruits histone deacetylase 3 (HDAC3) to suppress de novo lipogenesis and inhibit tumor metastasis in the pathogenesis of prostate cancer [45]. This is a good example on how a gene with high risk identified using genetic analysis allows revealing a new disease mechanism, which will hopefully lead to the development of a disease-specific drug such as a PARP inhibitor [46].

In pancreatic cancer, *BRCA2* ($P = 1.67 \times 10^{-9}$, OR = 9.8, 95% CI: 4.7–20.6), *ATM* ($P = 3.47 \times 10^{-4}$, OR = 6.5, 95% CI: 2.3–18.0), and *BRCA1* ($P = 6.09 \times 10^{-3}$,

Table 1 Carrier frequency of pathogenic variants in five cancer types in the Japanese population

	Female breast	Prostate	Pancreatic	Colorectal	Renal	Control
Sample	7091	7743	1009	12,578	849	38,161
Carrier frequency (%)						
APC	–	–	0.000	0.164	0.000	0.008
ATM	0.300	0.292	1.682	0.271	0.119	0.166
BARD1	–	–	0.099	0.036	0.000	0.053
BMPR1A	–	–	0.000	0.000	–	0.000
BRCA1	1.322	0.168	0.692	0.178	0.237	0.069
BRCA2	2.754	1.097	2.374	0.407	0.474	0.221
BRIP1	–	0.133	0.099	0.221	0.000	0.134
CDH1	0.020	–	0.000	0.021	0.000	0.013
CDK4	–	–	0.000	0.000	0.000	0.003
CDKN2A	–	–	0.099	0.007	0.000	0.008
CHEK2	0.250	0.186	0.099	0.136	0.119	0.090
EPCAM	–	–	0.099	0.029	0.000	0.021
HOXB13	–	0.009	0.000	0.021	0.000	0.024
MLH1	–	–	0.000	0.236	0.000	0.005
MSH2	–	–	0.099	0.335	0.000	0.021
MSH6	–	–	0.000	0.364	0.000	0.042
MUTYH	–	–	0.198	0.186	0.119	0.166
NBN	0.060	0.150	0.198	0.107	0.119	0.132
NF1	0.110	–	0.000	0.071	0.000	0.050
PALB2	0.371	0.080	0.297	0.100	0.000	0.053
PMS2	–	–	0.000	0.079	0.119	0.045
PTEN	0.130	–	0.099	0.021	0.000	0.021
RAD51C	–	–	0.000	0.029	0.119	0.037
RAD51D	–	–	0.198	0.314	0.119	0.324
SMAD4	–	–	0.000	0.000	0.000	0.008
STK11	0.000	–	0.000	0.007	0.000	0.005
TP53	0.130	–	0.099	0.057	0.000	0.016
Reference	[23]	[26]	[27]	[28]	[29]	

The hyphen indicates no analysis. In the present study, LoF variants or pathogenic variants registered in ClinVar were considered as pathogenic

OR = 7.6, 95% CI: 1.8–32.7) showed $P < 0.05$. In the original study [27], *APC*, *PALB2*, and *CDKN2A* showed $P < 0.05$; however, only one or two cases were reported. Thus, it might be difficult to conclude that these genes are also risk factors for pancreatic cancer. The disease risk of *BRCA1/2* and *ATM* was similar to that of the largest investigation in the European population [5]. Their frequency of patients with pancreatic cancer who were carriers of pathogenic variants in these three genes (5.1%) was similar to the frequency (4.8%) calculated in the present study.

In colorectal cancer, *APC* ($P = 1.02 \times 10^{-5}$, OR = 36.3, 95% CI: 7.4–178.8), *MLH1* ($P = 2.24 \times 10^{-5}$, OR = 33.6, 95% CI: 6.6–170.8), *MSH6* ($P = 2.69 \times 10^{-4}$, OR = 4.3, 95% CI: 2.0–9.4), *MSH2* ($P = 0.012$, OR = 4.3, 95% CI: 1.4–13.6), and

Table 2 Results of the association analysis of pathogenic variants between cases and controls showing $P < 0.05$

Gene	*P*	OR	95% CI lower	95% CI upper
Female breast cancer				
BRCA2	5.83E-24	8.7	5.7	13.3
BRCA1	4.98E-09	10.3	4.7	22.4
PTEN	8.56E-04	15.5	3.1	77.9
CHEK2	2.47E-03	3.5	1.6	7.8
PALB2	2.85E-03	4.0	1.6	9.8
Prostate cancer				
BRCA2	8.06E-08	4.2	2.5	7.1
Pancreatic cancer				
BRCA2	1.67E-09	9.8	4.7	20.6
ATM	3.47E-04	6.5	2.3	18.0
BRCA1	6.09E-03	7.6	1.8	32.7
Colrectal cancer				
APC	1.02E-05	36.3	7.4	178.8
MLH1	2.24E-05	33.6	6.6	170.8
MSH6	2.69E-04	4.3	2.0	9.4
MSH2	0.012	4.3	1.4	13.6
PTEN	0.048	4.0	1.0	15.5
Renal cancer				
PMS2	0.031	9.4	1.2	72.5

We reanalyzed the original data to directly compare the results of the five cancer types. For this purpose, LoF variants or pathogenic variants registered in ClinVar were considered as pathogenic, and we used only patients without a family history of cancer to provide a better estimation of disease risk, as we used controls without a family history of cancer. Therefore, our results were expected to differ from those of the original studies for each cancer type

PTEN ($P = 0.048$, OR = 4.0, 95% CI: 1.0–15.5) showed $P < 0.05$. When compared with the OR of *MLH1* in the original study [28] (OR = 8.6, 95% CI: 3.9–21.3), that of our study was much higher. In the present study, we used only cases without a family history of cancer to precisely calculate the OR, and therefore, we expected a lower OR than that in the original study. In fact, most of the results showed this tendency. However, *MLH1* showed the opposite trend. Although this should be carefully analyzed in future studies, the impact of a family history of cancer may vary between cancer types and genes because the family history includes different information on shared genetic and environmental factors [47].

In renal cancer, only *PMS2* showed $P < 0.05$. The original study [29] showed that the associated genes differed between clear and non-clear cell renal cell carcinoma (RCC). For the former, pathogenic variants in *TP53* ($P = 1.73 \times 10^{-4}$, OR = 5.8, 95% CI: 2.2–15.7), *CHEK2* ($P = 0.003$, OR = 7.0, 95% CI: 1.7–33.9), and *PMS2* ($P = 0.04$, OR = 7.0, 95% CI: 0.8–84.3) were associated with disease risk. In addition, because there are several causative genes of hereditary RCCs with specific histologic subtypes and clinical phenotypes, we analyzed an additional set of 13 genes (*BAP1, CDC73, FH, FLCN, MET, MITF, SDHA, SDHB, SDHC, SDHD, TSC1, TSC2*, and *VHL*). *PTEN*, a known causative gene of hereditary RCC, was already included in the 27 hereditary cancer genes. Among the 13 additional genes, *BAP1* ($P = 0.005$, OR = ∞, 95% CI: 1.9-∞) and *VHL* ($P = 0.03$, OR = 7.0, 95% CI: 0.8–84.3) showed disease risk association with clear cell RCC. For non-clear cell RCC, only hereditary RCC-specific genes showed an association with disease risk: *BAP1* ($P = 6.27 \times 10^{-5}$, OR = ∞, 95% CI: 10.0-∞), *FH* ($P = 6.27 \times 10^{-5}$, OR = ∞, 95% CI: 10.0-∞), *TSC1* ($P = 0.002$, OR = ∞, 95% CI: 4.5-∞), and *FLCN* ($P = 0.009$, OR = 24.2, 95% CI: 1.7–333.9). Note that none of the 5996 controls had pathogenic variants in all associated hereditary RCC-specific genes, except for *FLCN*. Although disease risk could not be calculated, a higher disease risk was considered based on a lower value of the 95% CI.

2.6 Variants of Uncertain Significance

Figure 2 shows the proportion of genetic variants within four categories in the five cancer types: (A) pathogenic variants (LoF variants or pathogenic variants in ClinVar); (B) benign variants in ClinVar; (C) missense variants whose annotation is VUS or not registered in ClinVar; and (D) synonymous variants whose annotation is VUS or not registered in ClinVar. (A) Pathogenic variants and (B) benign variants corresponded to 6.1%–13.5% and 18.8%–22.1%, respectively. Most of (D) synonymous variants are considered benign if they do not influence splicing [48, 49]. Although (C) is the most difficult to annotate and consider as VUS, they accounted for 55.6%–63.4% of all genetic variants and potentially include pathogenic variants. In our prostate cancer analysis [26], the gene-based association test using the VUS in *CHEK2* showed some association ($P = 8.45 \times 10^{-6}$, OR = 1.62, 95% CI: 1.30–2.20). The three variants, p.Ala496Pro (OR = 4.22, 95% CI: 1.41–15.11), p.Arg223Cys (OR = 1.98, 95% CI: 1.01–3.92), and p.His414Tyr (OR = 2.25, 95% CI: 1.04–4.99), showed $P < 0.05$. Based on only these association results, we could not conclude these variants were pathogenic; however, these results suggest that non-synonymous variants in *CHEK2* may contribute to an increase in disease risk.

Several functional analyses have been proposed for each gene to determine its pathogenesis, and *BRCA1/2* have been tested frequently. However, these genes have various diverse functions, such as homology-directed DNA repair function, embryonic stem cell viability, transcriptional activation, drug-sensitivity, protein–protein interaction, and splicing [50]. Some functional assays are reported to reflect the pathogenicity of variants registered in ClinVar [50, 51]. Disadvantages of these functional assays are that additional assays are needed as far as next-generation sequencers identify new genetic variants. In addition, different assays should be developed for different genes.

In silico assays have the potential to overcome these limitations. Several software packages have been developed, such as SIFT [52], Condel [53], and CADD [54], and some softwares can annotate genetic variants outside the coding regions.

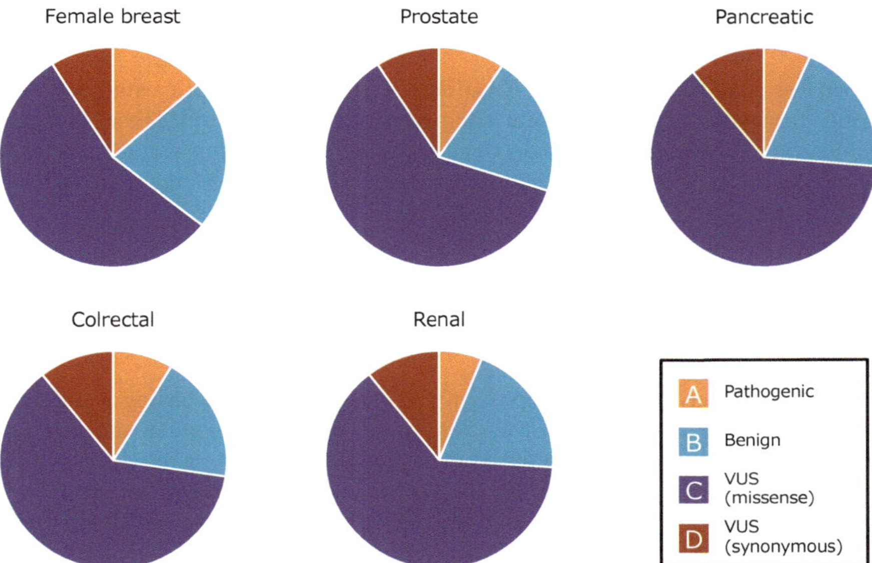

Fig. 2 The proportion of genetic variants in each of the four categories for the five cancer types. (**a**) Pathogenic variants (LoF variants or pathogenic variants in ClinVar); (**b**) benign variants in ClinVar; (**c**) missense variants whose annotation is VUS or not registered in ClinVar; and (**d**) synonymous variants whose annotation is VUS or not registered in ClinVar

Comparison studies of these performance have been conducted [55]. Recently, an evolutionary model of variant effect outperformed other computational approaches in terms of concordance with the pathogenicity registered in ClinVar [56].

However, "calibration" is important in both in vitro and in silico methods. The proportion of functional reduction leading to increased disease risk should depend on the combination of a gene with a disease. In a homology-directed DNA repair assay in *BRCA2*, ROC curve analysis of the normalized mean functional assay results for the established pathogenic and neutral variants identified their thresholds [57]. In addition, although a borderline reduction might increase the disease risk, this risk might be lower than that of LoF variants. For instance, R1699Q in *BRCA1* has an estimated cumulative risk to breast or ovarian cancer of 24% at 70 years of age [58], which could provide a less-intensive clinical option to patients. A "calibration" is therefore very important for translating results from in vivo or in vitro assays of VUS for clinical use. One robust method is that each assay with a certain threshold for pathogenicity should test new genetic data with clinical information and show the enrichment of pathogenic variants annotated by them in case samples as compared with control samples. Their estimated risk should be equal to the risk of the LoF variants.

3 Heterogeneity

When we compared the results of various cancer types in Japan with published results from other countries, several heterogeneities were observed. These heterogeneities should be thoroughly understood in future studies using various methods. These efforts could provide additional evidence for improving personalized medicine.

3.1 Between Genes

Heterogeneity between genes appears to be due to each gene having a different function. Although *BRCA1* and *BRCA2* are known to have both similar and different functions [59], they are often referred to as *BRCA1/2* and are dealt with together. However, these different functions should be carefully considered in clinical settings. A well-known example is ovarian cancer. The cumulative risks of ovarian cancer are low up to 40 years for *BRCA1* carriers and age 50 years for *BRCA2* carriers in the European populations [60]. This information is of utmost importance for considering risk-reducing salpingo-oophorectomy for a female carrier of a pathogenic variant in *BRCA1* or *BRCA2*. Another example is the difference in prostate cancer risk between pathogenic variants of *BRCA1* and *BRCA2*, as described above. The difference in disease risk between both genes might lead to a different therapeutic effect of a PARP inhibitor for prostate cancer. However, this difference is sometimes not considered in the discussion of *BRAC1/2* clinical uses [61].

In a broader context, *BRCA1* and *BRCA2* are involved in homologous recombination and several other associated genes could be hypothesized to have a similar impact in the clinics. For instance, a pathogenic variant of these genes might confer a high risk of breast and other *BRCA1/2*-associated cancer types, and a carrier patient with a pathogenic variant may have a better therapeutic effect of a PARP inhibitor. Based on this hypothesis, several candidate genes studies have been analyzed. One of the most successful outcomes is the identification of *PALB2* because it interacts with *BRCA2*, and biallelic pathogenic variants in *PALB2*, similar to biallelic *BRCA2* variants, cause Fanconi anemia [62]. However, the disease risk is different from those of *BRCA1* and *BRCA2* [25]. For geneticists, it would be reasonable to identify many candidate gene studies before the GWAS era, but most of those studies failed to be replicated [63]. Therefore, each gene in each cancer type should be carefully examined in a large-scale sample for precise clinical use.

3.2 Between Cancer Types

The cancer type targeted by a pathogenic variant in each gene is important for the surveillance of a pathogenic variant carrier. However, it seems impossible to identify the targeted cancer type based on the knowledge of gene function and organs. This has been examined in around 2000 for *BRCA1* and *BRCA2*. For instance, a cohort study of 11,847 individuals from 699 families with a *BRCA1* pathogenic variant was examined, and the observed cancer incidence was compared with the expected cancer incidence based on population cancer rates [9]. The relative risks of pancreatic cancer (2.26), uterine body cancer (2.65), and cervical cancer (3.72) were identified in 2001. A similar analysis of *BRCA2* was reported in 1999 [7]. Various reports have been published, but only breast, ovarian, prostate, and pancreatic cancers were considered to have established BRCA1/2-targeted genes in 2021 because of limited evidence for other cancer types.

In the preparation of this study, our group reported a large-scale genetic analysis of *BRCA1/2* across 14 cancer types in 65,108 patients and 38,153 controls [41]. We identified pathogenic variants associated with an OR $\geq$ 4 in biliary tract cancer (OR = 17.4) in *BRCA1*, esophageal cancer (OR = 5.6) in *BRCA2*, and gastric cancer in *BRCA1* (OR = 5.2) and *BRCA2* (OR = 4.7), in addition to the four established cancer types (i.e., breast, ovarian, prostate, and pancreatic). Among them, the cumulative absolute risk at 85 years of age was estimated to be as high as 20% in gastric cancer. Biliary, esophageal, and gastric cancers are more common in East Asia [22]. Carrier patients of these genes with pathogenic variants are expected to have a certain therapeutic effect of a PARP inhibitor; however, additional clinical trials are needed for clinical use. These results highlight the importance of large-scale genetic analyses of each population.

A study with a purpose similar to ours, but using a different approach was published at the same time. The researchers analyzed data from 3184 *BRCA1* and 2157 *BRCA2* carrier families to estimate disease risk of the 22 primary cancer types, adjusting for family ascertainment in mainly European populations [64]. Although they also identified the relative risk of gastric cancer in *BRCA1* (2.17) and *BRCA2* (3.69), their cumulative risk at 80 years of age was much lower (0.7–3.5%). This difference from the Japanese data could be partially explained by the difference in the prevalence of gastric cancer [22].

In addition, the phenotype-wide association study using health records from 214,020 participants of three large-scale cohorts namely, the Electronic Medical Records and Genomics Sequencing data set, the UK Biobank cohort, and the Hereditary Cancer Registry identified that pathogenic variants in hereditary cancer genes were associated with not only neoplasms but also non-neoplastic diseases [65]. For example, pathogenic variants were associated with ovarian cysts in *BRCA1* (OR = 3.2) and *BRCA2* (OR = 3.1). These three recent studies highlight that genetic analysis using very large sample sizes could help identify the target cancer types of hereditary cancer genes.

3.3 Between Populations

Generally, the association between genes and cancer types is consistent among populations. However, founder pathogenic variants and the prevalence of each cancer type can change the clinical utility of their genes in each population. For instance, *NBN* is one of the targeted genes for genetic testing of breast cancer. However, there is no association between *NBN* and breast cancer reported in Japan [23]. One reason could be that only 1/7051 female breast cancer patients has a pathogenic variant of *NBN*. However, c.657del5 (rs587776650) of *NBN* is a founder pathogenic variant in the Polish population [66], where it is common and contributes to an increased risk of both breast [66] and prostate cancer [67]. Furthermore, this variant was not observed in Japanese patients, and only another pathogenic variant was observed. Therefore, the current results suggest that the clinical utility of *NBN* is very limited in Japanese patients with breast cancer.

Another interesting example is *HOXB13* in prostate cancer. This gene showed an association in both European and Japanese populations. However, the key founder pathogenic variant was different: p.Gly84Glu in the European population [43] and p.Gly132Glu in the Japanese population. Recently, it was revealed how a pathogenic variant in *HOXB13* increases the disease risk of prostate cancer, but the difference in amino acid positions was not examined [26]. In *BRCA1/2*, amino acid positions of pathogenic variants influence breast and ovarian cancer risks [68]. Therefore, such differences may cause population differences in the clinical characteristics of prostate cancer caused by different pathogenic variants of *HOXB13*.

4 Future Directions of Research

4.1 Integrated Estimation of Disease Risk with Genetic and Environmental Factors

Some of these data have already been described in the Guidelines for Diagnosis and Treatment of Hereditary Breast and Ovarian Cancer 2021 as contributors of personalized medicine in Japan. However, there is room for improvement. Here, we present two directions for future research.

We analyzed hereditary cancer genes with a high risk, but they play a role in a part of genetic effects on cancer onset. Other effects from genetic variants with relatively small risk (<1.2) should also be considered [20]. We could extract the effects of common variants (population frequency > 1%) as polygenic risk scores [69]. Some patients with pathogenic variants in *BRCA2* develop breast cancer first, whereas others develop ovarian cancer first. The first cancer in a carrier is different, but this might be partially explained by the different polygenic scores between *BRCA2*-related cancers. In the future, hereditary cancer genes and polygenic risk scores in an individual could indicate which cancer type should be more cared [70].

Several factors, such as smoking, alcohol consumption, food intake, obesity, reproductive history, hormone intake, and bacterial and viral infections, were known to influence cancer types differently [71]. A better understanding of these factors could contribute to improve personalized medicine.

4.2 *Identification of More Genes with Moderate or High Risk*

More genes might have clinical potential for personalized medicine, as most genes were not yet examined by genome-wide approach. *BRCA1* and *BRCA2* were identified by one of genome-wide approach, linkage analysis [2, 3]. However, linkage analysis requires many families, each of which including multiple patients and controls. Currently, it is difficult to collect data on such families. GWAS [72] was used to identify a few hundred genome loci associated with various cancer types [73, 74]. The advantage of this method is the collection of patients and controls without family information, and a meta-analysis combining several GWAS could easily increase the sample size to identify more associated variants [72]. However, this approach can only analyze common genetic variants with a very low impact (i.e., OR < 1.2). A recent GWAS with a large sample size and imputation data could identify founder pathogenic variants with high risk but not pathogenic variants with very low frequency because the reference panel for imputation does not include haplotypes with very rare pathogenic variants.

WGS can identify all the genetic variants. A key difficulty is that much more samples are required for rare variant analysis because the frequency of rare variants is very low (<0.1% or lower). For instance, when OR = ~1.2, and the minor allele frequency of the variant is 0.1%, one million individuals are required to obtain genome-wide significance at $P = 5 \times 10^{-8}$ with 80% statistical power. Although pathogenic variants are expected to have a higher disease risk, a large number of samples are still needed. To overcome this issue, a burden test combining rare variants in each functional unit, normally a gene, could increase its statistical power [20]. However, several thousand samples would not be sufficient for this purpose, according to the results from a recent exome-sequencing approach in the UK Biobank [75]. *BRCA2* is definitely the gene responsible for prostate cancer but shows only $P < 1 \times 10^{-4}$ in 2875 cases and 148,067 controls. This association is nearly impossible to identify among many genes with similar *P* values. One potential route is to perform target sequencing of genes identified by GWAS. Recent WGS studies have identified rare new variants in genes previously identified using GWAS [76]. Although it is challenging to identify target genes from associated variants in GWAS, various bioinformatics pipelines using deposited omics data on expression and methylation for each cell type are being developed for this purpose [77]. By focusing on specific reasonable positional-candidate genes instead of the whole genome, target sequencing could be used to analyze much more samples to increase statistical power. This method has identified additional rare variants in various diseases [20].

5 Conclusions

BRCA1/2 is considered a top runner in personalized medicine. However, this information cannot be used in clinics without concern. We reanalyzed genetic data on breast, prostate, pancreatic, colorectal, and renal cancers to better compare these cancer types in the Japanese population. The various heterogeneities were observed, and future research directions were then discussed. Further research using pathogenic germline variants could improve personalized medicine and provide lessons for personalized medicine in other diseases.

References

1. Armstrong K, Weiner J, Weber B, Asch DA. Early adoption of BRCA1/2 testing: who and why. Genet Med. 2003;5(2):92–8.
2. Wooster R, Bignell G, Lancaster J, Swift S, Seal S, Mangion J, et al. Identification of the breast cancer susceptibility gene BRCA2. Nature. 1995;378(6559):789–92.
3. Miki Y, Swensen J, Shattuck-Eidens D, Futreal PA, Harshman K, Tavtigian S, et al. A strong candidate for the breast and ovarian cancer susceptibility gene BRCA1. Science. 1994;266(5182):66–71.
4. Oh M, Alkhushaym N, Fallatah S, Althagafi A, Aljadeed R, Alsowaida Y, et al. The association of BRCA1 and BRCA2 mutations with prostate cancer risk, frequency, and mortality: a meta-analysis. Prostate. 2019;79(8):880–95.
5. Hu C, Hart SN, Polley EC, Gnanaolivu R, Shimelis H, Lee KY, et al. Association between inherited germline mutations in cancer predisposition genes and risk of pancreatic cancer. JAMA. 2018;319(23):2401–9.
6. Jonsson P, Bandlamudi C, Cheng ML, Srinivasan P, Chavan SS, Friedman ND, et al. Tumour lineage shapes BRCA-mediated phenotypes. Nature. 2019;571(7766):576–9.
7. Breast Cancer Linkage C. Cancer risks in BRCA2 mutation carriers. J Natl Cancer Inst. 1999;91(15):1310–6.
8. Johannsson O, Loman N, Moller T, Kristoffersson U, Borg A, Olsson H. Incidence of malignant tumours in relatives of BRCA1 and BRCA2 germline mutation carriers. Eur J Cancer. 1999;35(8):1248–57.
9. Thompson D, Easton DF, Breast Cancer Linkage C. Cancer incidence in BRCA1 mutation carriers. J Natl Cancer Inst. 2002;94(18):1358–65.
10. Ko JM, Ning L, Zhao XK, Chai AWY, Lei LC, Choi SSA, et al. BRCA2 loss-of-function germline mutations are associated with esophageal squamous cell carcinoma risk in Chinese. Int J Cancer. 2020;146(4):1042–51.
11. Moran A, O'Hara C, Khan S, Shack L, Woodward E, Maher ER, et al. Risk of cancer other than breast or ovarian in individuals with BRCA1 and BRCA2 mutations. Familial Cancer. 2012;11(2):235–42.
12. Jakubowska A, Nej K, Huzarski T, Scott RJ, Lubinski J. BRCA2 gene mutations in families with aggregations of breast and stomach cancers. Br J Cancer. 2002;87(8):888–91.
13. Richards S, Aziz N, Bale S, Bick D, Das S, Gastier-Foster J, et al. Standards and guidelines for the interpretation of sequence variants: a joint consensus recommendation of the American College of Medical Genetics and Genomics and the Association for Molecular Pathology. Genet Med. 2015;17(5):405–24.
14. Spurdle AB, Healey S, Devereau A, Hogervorst FB, Monteiro AN, Nathanson KL, et al. ENIGMA—Evidence-based network for the interpretation of germline mutant alleles: an

international initiative to evaluate risk and clinical significance associated with sequence variation in BRCA1 and BRCA2 genes. Hum Mutat. 2012;33(1):2–7.
15. Fanale D, Pivetti A, Cancelliere D, Spera A, Bono M, Fiorino A, et al. BRCA1/2 variants of unknown significance in hereditary breast and ovarian cancer (HBOC) syndrome: looking for the hidden meaning. Crit Rev Oncol Hematol. 2022;172:103626.
16. Casaletto J, Parsons M, Markello C, Iwasaki Y, Momozawa Y, Spurdle AB, et al. Federated analysis of BRCA1 and BRCA2 variation in a Japanese cohort. Cell Genom. 2022;2(3):100109.
17. Rebbeck TR, Friebel TM, Friedman E, Hamann U, Huo D, Kwong A, et al. Mutational spectrum in a worldwide study of 29,700 families with BRCA1 or BRCA2 mutations. Hum Mutat. 2018;39(5):593–620.
18. Manchanda R, Lieberman S, Gaba F, Lahad A, Levy-Lahad E. Population screening for inherited predisposition to breast and ovarian cancer. Annu Rev Genomics Hum Genet. 2020;21:373–412.
19. Easton DF, Pharoah PD, Antoniou AC, Tischkowitz M, Tavtigian SV, Nathanson KL, et al. Gene-panel sequencing and the prediction of breast-cancer risk. N Engl J Med. 2015;372(23):2243–57.
20. Momozawa Y, Mizukami K. Unique roles of rare variants in the genetics of complex diseases in humans. J Hum Genet. 2021;66(1):11–23.
21. Duggan C, Trapani D, Ilbawi AM, Fidarova E, Laversanne M, Curigliano G, et al. National health system characteristics, breast cancer stage at diagnosis, and breast cancer mortality: a population-based analysis. Lancet Oncol. 2021;22(11):1632–42.
22. Sung H, Ferlay J, Siegel RL, Laversanne M, Soerjomataram I, Jemal A, et al. Global cancer statistics 2020: GLOBOCAN estimates of incidence and mortality worldwide for 36 cancers in 185 countries. CA Cancer J Clin. 2021;71(3):209–49.
23. Momozawa Y, Iwasaki Y, Parsons MT, Kamatani Y, Takahashi A, Tamura C, et al. Germline pathogenic variants of 11 breast cancer genes in 7,051 Japanese patients and 11,241 controls. Nat Commun. 2018;9(1):4083.
24. Hu C, Hart SN, Gnanaolivu R, Huang H, Lee KY, Na J, et al. A population-based study of genes previously implicated in breast cancer. N Engl J Med. 2021;384(5):440–51.
25. Breast Cancer Association C, Dorling L, Carvalho S, Allen J, Gonzalez-Neira A, Luccarini C, et al. Breast cancer risk genes - association analysis in more than 113,000 women. N Engl J Med. 2021;384(5):428–39.
26. Momozawa Y, Iwasaki Y, Hirata M, Liu X, Kamatani Y, Takahashi A, et al. Germline pathogenic variants in 7,636 Japanese patients with prostate cancer and 12,366 controls. J Natl Cancer Inst. 2020;112(4):369–76.
27. Mizukami K, Iwasaki Y, Kawakami E, Hirata M, Kamatani Y, Matsuda K, et al. Genetic characterization of pancreatic cancer patients and prediction of carrier status of germline pathogenic variants in cancer-predisposing genes. EBioMedicine. 2020;60:103033.
28. Fujita M, Liu X, Iwasaki Y, Terao C, Mizukami K, Kawakami E, et al. Population-based screening for hereditary colorectal cancer variants in Japan. Clin Gastroenterol Hepatol. 2022;20(9):2132–2141.e9.
29. Sekine Y, Iwasaki Y, Aoi T, Endo M, Hirata M, Kamatani Y, et al. Different risk genes contribute to clear cell and non-clear cell renal cell carcinoma in 1532 Japanese patients and 5996 controls. Hum Mol Genet. 2022;31(12):1962–9.
30. Hirata M, Kamatani Y, Nagai A, Kiyohara Y, Ninomiya T, Tamakoshi A, et al. Cross-sectional analysis of BioBank Japan clinical data: a large cohort of 200,000 patients with 47 common diseases. J Epidemiol. 2017;27(3S):S9–S21.
31. Nagai A, Hirata M, Kamatani Y, Muto K, Matsuda K, Kiyohara Y, et al. Overview of the BioBank Japan project: study design and profile. J Epidemiol. 2017;27(3S):S2–8.
32. Hirata M, Nagai A, Kamatani Y, Ninomiya T, Tamakoshi A, Yamagata Z, et al. Overview of BioBank Japan follow-up data in 32 diseases. J Epidemiol. 2017;27(3S):S22–S8.

33. Ishigaki K, Akiyama M, Kanai M, Takahashi A, Kawakami E, Sugishita H, et al. Large-scale genome-wide association study in a Japanese population identifies novel susceptibility loci across different diseases. Nat Genet. 2020;52(7):669–79.
34. Sakaue S, Kanai M, Tanigawa Y, Karjalainen J, Kurki M, Koshiba S, et al. A cross-population atlas of genetic associations for 220 human phenotypes. Nat Genet. 2021;53(10):1415–24.
35. Karczewski KJ, Francioli LC, Tiao G, Cummings BB, Alfoldi J, Wang Q, et al. The mutational constraint spectrum quantified from variation in 141,456 humans. Nature. 2020;581(7809):434–43.
36. Momozawa Y, Akiyama M, Kamatani Y, Arakawa S, Yasuda M, Yoshida S, et al. Low-frequency coding variants in CETP and CFB are associated with susceptibility of exudative age-related macular degeneration in the Japanese population. Hum Mol Genet. 2016;25(22):5027–34.
37. Landrum MJ, Lee JM, Benson M, Brown G, Chao C, Chitipiralla S, et al. ClinVar: public archive of interpretations of clinically relevant variants. Nucleic Acids Res. 2016;44(D1):D862–8.
38. Amendola LM, Jarvik GP, Leo MC, McLaughlin HM, Akkari Y, Amaral MD, et al. Performance of ACMG-AMP variant-interpretation guidelines among nine Laboratories in the Clinical Sequencing Exploratory Research Consortium. Am J Hum Genet. 2016;98(6):1067–76.
39. Tayoun AN, Pesaran T, DiStefano MT, Oza A, Rehm HL, Biesecker LG, et al. Recommendations for interpreting the loss of function PVS1 ACMG/AMP variant criterion. Hum Mutat. 2018;39(11):1517–24.
40. Mighton C, Charames GS, Wang M, Zakoor KR, Wong A, Shickh S, et al. Variant classification changes over time in BRCA1 and BRCA2. Genet Med. 2019;21(10):2248–54.
41. Momozawa Y, Sasai R, Usui Y, Shiraishi K, Iwasaki Y, Taniyama Y, et al. Expansion of cancer risk profile for BRCA1 and BRCA2 pathogenic variants. JAMA Oncol. 2022;8(6):871–8.
42. Wei Y, Wu J, Gu W, Qin X, Dai B, Lin G, et al. Germline DNA repair gene mutation landscape in Chinese prostate cancer patients. Eur Urol. 2019;76(3):280–3.
43. Ewing CM, Ray AM, Lange EM, Zuhlke KA, Robbins CM, Tembe WD, et al. Germline mutations in HOXB13 and prostate-cancer risk. N Engl J Med. 2012;366(2):141–9.
44. Lin X, Qu L, Chen Z, Xu C, Ye D, Shao Q, et al. A novel germline mutation in HOXB13 is associated with prostate cancer risk in Chinese men. Prostate. 2013;73(2):169–75.
45. Lu X, Fong KW, Gritsina G, Wang F, Baca SC, Brea LT, et al. HOXB13 suppresses de novo lipogenesis through HDAC3-mediated epigenetic reprogramming in prostate cancer. Nat Genet. 2022;54(5):670–83.
46. Curtin NJ, Szabo C. Poly(ADP-ribose) polymerase inhibition: past, present and future. Nat Rev Drug Discov. 2020;19(10):711–36.
47. Valdez R, Yoon PW, Qureshi N, Green RF, Khoury MJ. Family history in public health practice: a genomic tool for disease prevention and health promotion. Annu Rev Public Health. 2010;31:69–87.
48. Claes K, Poppe B, Machackova E, Coene I, Foretova L, De Paepe A, et al. Differentiating pathogenic mutations from polymorphic alterations in the splice sites of BRCA1 and BRCA2. Genes Chromosomes Cancer. 2003;37(3):314–20.
49. Colombo M, De Vecchi G, Caleca L, Foglia C, Ripamonti CB, Ficarazzi F, et al. Comparative in vitro and in silico analyses of variants in splicing regions of BRCA1 and BRCA2 genes and characterization of novel pathogenic mutations. PLoS One. 2013;8(2):e57173.
50. Findlay GM, Daza RM, Martin B, Zhang MD, Leith AP, Gasperini M, et al. Accurate classification of BRCA1 variants with saturation genome editing. Nature. 2018;562(7726):217–22.
51. Ikegami M, Kohsaka S, Ueno T, Momozawa Y, Inoue S, Tamura K, et al. High-throughput functional evaluation of BRCA2 variants of unknown significance. Nat Commun. 2020;11(1):2573.
52. Kumar P, Henikoff S, Ng PC. Predicting the effects of coding non-synonymous variants on protein function using the SIFT algorithm. Nat Protoc. 2009;4(7):1073–81.
53. Gonzalez-Perez A, Lopez-Bigas N. Improving the assessment of the outcome of non-synonymous SNVs with a consensus deleteriousness score, Condel. Am J Hum Genet. 2011;88(4):440–9.

54. Rentzsch P, Witten D, Cooper GM, Shendure J, Kircher M. CADD: predicting the deleteriousness of variants throughout the human genome. Nucleic Acids Res. 2019;47(D1):D886–D94.
55. Ernst C, Hahnen E, Engel C, Nothnagel M, Weber J, Schmutzler RK, et al. Performance of in silico prediction tools for the classification of rare BRCA1/2 missense variants in clinical diagnostics. BMC Med Genet. 2018;11(1):35.
56. Frazer J, Notin P, Dias M, Gomez A, Min JK, Brock K, et al. Disease variant prediction with deep generative models of evolutionary data. Nature. 2021;599(7883):91–5.
57. Guidugli L, Shimelis H, Masica DL, Pankratz VS, Lipton GB, Singh N, et al. Assessment of the clinical relevance of BRCA2 missense variants by functional and computational approaches. Am J Hum Genet. 2018;102(2):233–48.
58. Spurdle AB, Whiley PJ, Thompson B, Feng B, Healey S, Brown MA, et al. BRCA1 R1699Q variant displaying ambiguous functional abrogation confers intermediate breast and ovarian cancer risk. J Med Genet. 2012;49(8):525–32.
59. Prakash R, Zhang Y, Feng W, Jasin M. Homologous recombination and human health: the roles of BRCA1, BRCA2, and associated proteins. Cold Spring Harb Perspect Biol. 2015;7(4):a016600.
60. Kuchenbaecker KB, Hopper JL, Barnes DR, Phillips K-A, Mooij TM, Roos-Blom M-J, et al. Risks of breast, ovarian, and contralateral breast cancer for BRCA1 and BRCA2 mutation carriers. JAMA. 2017;317(23):2402–16.
61. Teyssonneau D, Margot H, Cabart M, Anonnay M, Sargos P, Vuong NS, et al. Prostate cancer and PARP inhibitors: progress and challenges. J Hematol Oncol. 2021;14(1):51.
62. Rahman N, Seal S, Thompson D, Kelly P, Renwick A, Elliott A, et al. PALB2, which encodes a BRCA2-interacting protein, is a breast cancer susceptibility gene. Nat Genet. 2007;39(2):165–7.
63. Tabor HK, Risch NJ, Myers RM. Opinion: candidate-gene approaches for studying complex genetic traits: practical considerations. Nat Rev Genet. 2002;3(5):391–7.
64. Li S, Silvestri V, Leslie G, Rebbeck TR, Neuhausen SL, Hopper JL, et al. Cancer risks associated with BRCA1 and BRCA2 pathogenic variants. J Clin Oncol. 2022;40(14):1529–41.
65. Zeng C, Bastarache LA, Tao R, Venner E, Hebbring S, Andujar JD, et al. Association of pathogenic variants in hereditary cancer genes with multiple diseases. JAMA Oncol. 2022;8(6):835–44.
66. Zhang G, Zeng Y, Liu Z, Wei W. Significant association between Nijmegen breakage syndrome 1 657del5 polymorphism and breast cancer risk. Tumour Biol. 2013;34(5):2753–7.
67. Cybulski C, Wokołorczyk D, Kluźniak W, Jakubowska A, Górski B, Gronwald J, et al. An inherited NBN mutation is associated with poor prognosis prostate cancer. Br J Cancer. 2012;108(2):461–8.
68. Rebbeck TR, Mitra N, Wan F, Sinilnikova OM, Healey S, McGuffog L, et al. Association of type and location of BRCA1 and BRCA2 mutations with risk of breast and ovarian cancer. JAMA. 2015;313(13):1347–61.
69. Choi SW, Mak TS, O'Reilly PF. Tutorial: a guide to performing polygenic risk score analyses. Nat Protoc. 2020;15(9):2759–72.
70. Gao C, Polley EC, Hart SN, Huang H, Hu C, Gnanaolivu R, et al. Risk of breast cancer among carriers of pathogenic variants in breast cancer predisposition genes varies by polygenic risk score. J Clin Oncol. 2021;39(23):2564–73.
71. Bruno E, Oliverio A, Paradiso A, Daniele A, Tommasi S, Terribile DA, et al. Lifestyle characteristics in women carriers of BRCA mutations: results from an Italian trial cohort. Clin Breast Cancer. 2021;21(3):e168–e76.
72. Tam V, Patel N, Turcotte M, Bossé Y, Paré G, Meyre D. Benefits and limitations of genome-wide association studies. Nat Rev Genet. 2019;20(8):467–84.
73. Zhang H, Ahearn TU, Lecarpentier J, Barnes D, Beesley J, Qi G, et al. Genome-wide association study identifies 32 novel breast cancer susceptibility loci from overall and subtype-specific analyses. Nat Genet. 2020;52(6):572–81.

74. Conti DV, Darst BF, Moss LC, Saunders EJ, Sheng X, Chou A, et al. Trans-ancestry genome-wide association meta-analysis of prostate cancer identifies new susceptibility loci and informs genetic risk prediction. Nat Genet. 2021;53(1):65–75.
75. Wang Q, Dhindsa RS, Carss K, Harper AR, Nag A, Tachmazidou I, et al. Rare variant contribution to human disease in 281,104 UK Biobank exomes. Nature. 2021;597(7877):527–32.
76. Luo Y, de Lange KM, Jostins L, Moutsianas L, Randall J, Kennedy NA, et al. Exploring the genetic architecture of inflammatory bowel disease by whole-genome sequencing identifies association at ADCY7. Nat Genet. 2017;49(2):186–92.
77. Mountjoy E, Schmidt EM, Carmona M, Schwartzentruber J, Peat G, Miranda A, et al. An open approach to systematically prioritize causal variants and genes at all published human GWAS trait-associated loci. Nat Genet. 2021;53(11):1527–33.

New Functions of *BRCA1/2* in Regulating Carcinogenesis and Drug Sensitivity

Zhenzhou Fang, Yuki Yoshino, and Natsuko Chiba

Abstract Pathogenic variants in *BRCA1* and *BRCA2* (*BRCA1/2*) cause hereditary breast and ovarian cancer syndrome (HBOC). BRCA1/2 are involved in multiple cellular processes, including homologous recombination (HR), which maintain genome stability. The HR repair function of BRCA1/2 is thought to underpin their function as tumor suppressors. However, it is unclear how dysregulation of these proteins causes tissue-specific carcinogenesis. Previously, we found that cancer-derived variants and abnormal expression of BRCA1-associated proteins cause centrosome amplification in mammary tissue-derived cells, resulting in chromosome segregation errors.

HR-deficient cells are sensitive to poly (ADP-ribose) polymerase inhibitors and platinum agents. Recently, we developed an HR activity assay named Assay of Site-Specific HR Activity (ASHRA), which evaluates HR activity quantitively. Analyzing the HR activity of BRCA1 variants using ASHRA revealed that the assay can predict whether an individual has a moderate risk of breast and ovarian cancer, and their sensitivity to PARP inhibitors. Furthermore, we identified a novel mechanism underlying resistance to the PARP inhibitor olaparib and the platinum agent cisplatin, which is dependent on high expression of activating transcription factor 1 (ATF1) and the transactivation activity of BRCA1 with ATF1.

In this chapter, we describe the effects of BRCA1/2 impairment, which is thought to contribute to carcinogenesis, as well as regulation of centrosome number by

Z. Fang
Department of Cancer Biology, Institute of Development, Aging and Cancer (IDAC), Tohoku University, Sendai, Japan

Y. Yoshino · N. Chiba (✉)
Department of Cancer Biology, Institute of Development, Aging and Cancer (IDAC), Tohoku University, Sendai, Japan

Laboratory of Cancer Biology, Graduate School of Life Sciences, Tohoku University, Sendai, Japan

Department of Cancer Biology, Tohoku University Graduate School of Medicine, Sendai, Japan
e-mail: natsuko.chiba.c7@tohoku.ac.jp

D. Aoki et al. (eds.), *Practical Guide to Hereditary Breast and Ovarian Cancer*,
https://doi.org/10.1007/978-981-99-5231-1_7

BRCA1, which may play a role in tissue-specific carcinogenesis. Furthermore, we describe the mechanisms underlying resistance to PARP inhibitors and suggest a novel mechanism by which BRCA1/ATF1-mediated transcription leads to resistance to olaparib and cisplatin.

Keywords Hereditary breast and ovarian cancer (HBOC) · *BRCA1/2* · Carcinogenesis · Drug sensitivity · PARP inhibitors

1 Introduction

Hereditary breast and ovarian cancer syndrome (HBOC) is a *BRCA1* and *BRCA2* *(BRCA1/2)*-linked genetic disorder associated with a high risk of breast, ovarian, and other cancers [1–3]. A diagnosis of HBOC is made by genetic testing for *BRCA1/2.* Clinical management approaches include screening for early cancer detection, prophylactic surgery for healthy carriers, and chemotherapy for patients with cancer. Therefore, accurate diagnoses of pathogenic variants are critical for clinical decision-making and improved prognosis. The pathogenicity of *BRCA1/2* variants is classified as benign, likely benign, variants of uncertain significance (VUS), likely pathogenic, or pathogenic. Most pathogenic variants are premature truncation variants generated by nonsense or frameshift mutations, whereas VUS are missense, small in-frame deletion or insertion, and splicing variants. The effects of these variants on function have not yet been determined.

BRCA1/2 functions in multiple cellular processes to maintain genome stability [4–6]. Homologous recombination (HR) repair is a critical error-free pathway for repairing DNA double-strand breaks; this pathway uses an intact sister chromatid as a template. BRCA1/2 function in HR repair is thought to underpin their role as tumor suppressors. Therefore, assays that evaluate HR activity have been used to estimate the pathogenicity of *BRCA1/2* variants. Of these, the direct-repeat GFP (DR-GFP) assay is the most common [7–13].

On the other hand, it is unclear how dysregulation of BRCA1/2 causes tissue-specific carcinogenesis. Centrosomes, the major microtubule nucleation centers in animal cells, mediate formation of a bipolar spindle during mitosis [14]. Aberration of centrosome number and structure are common in various cancers [15] and are related to an invasive phenotype [16]. Centrosome amplification results in chromosome segregation errors, leading to chromosomal instability, which is in turn associated with carcinogenesis and cancer progression [17]. Recently, we identified BRCA1-interacting proteins that function to regulate centrosomes together with BRCA1. We found that abnormalities in these BRCA1-associated proteins in mammary tissue-derived cells cause centrosome amplification [18–20].

In addition to its role in carcinogenesis, HR repair activity in cells is important for predicting sensitivity to some anti-cancer agents, poly (ADP-ribose) polymerase (PARP) inhibitors, and platinum agents. PARP functions in various DNA damage repair pathways as well as the repair process for DNA single-stranded breaks (SSBs) [21]. PARP inhibitors impair the repair of DNA single-stranded breaks (SSBs), which results in the creation of DNA double-stranded breaks (DSBs), and trap PARP protein at DNA SSBs [22]. Furthermore, because PARP helps to restart stalled DNA replication forks, PARP inhibitors induce collapse of replication forks [23, 24]. Since HR contributes to repair of DSBs, PARP trapping, and collapse of replication forks [22], PARP inhibitors cause synthetic lethality in cells with HR deficiency caused by alteration of HR factors, including BRCA1/2 [25–28].

PARP inhibitors have been developed to treat various cancers, including breast, ovarian, pancreatic, and prostate cancers [29]; however, a number of resistance mechanisms have been reported [30–32]. Recently, we developed an assay to evaluate HR activity in cells and analyzed the HR activity of BRCA1 variants [33]. We found that this assay evaluates HR activity quantitively [34]. We also identified a novel mechanism underlying resistance via transactivation activity of BRCA1 [34].

Here, we describe the functions of BRCA1/2, whose impairment is involved in carcinogenesis. We also discuss our recent finding that BRCA1-interacting proteins regulate centrosomes. Furthermore, we explain the mechanisms underlying resistance to PARP inhibitors and describe a novel mechanism of resistance to PARP inhibitors and platinum agents that is dependent on the function of BRCA1 during transcription.

2 Structure of BRCA1/2

BRCA1 has a RING domain at its amino (N)-terminal region that binds directly to BARD1. There is a coiled-coil motif and two BRCT domains at the carboxy (C)-terminal region (Fig. 1a). The coiled-coil motif binds to PALB2, and the BRCT domains bind to BRIP1, CtIP, and ABRAXAS1. *BARD1*, *PALB2*, *BRIP1*, and *ABRAXAS1* are also breast cancer susceptibility genes [6, 35].

BRCA2 has eight BRC motifs in its middle segment. BRCA2 binds to the RAD51 recombinase (which plays a critical role in HR activity) via BRC motifs and its C-terminal region (Fig. 1b). The N-terminal region contains a PALB2 binding domain. The C-terminal region contains a helical domain and three oligonucleotide/oligosaccharide binding (OB) folds that bind to ssDNA. The N-terminal DNA binding domain (NTD) also has DNA binding activity.

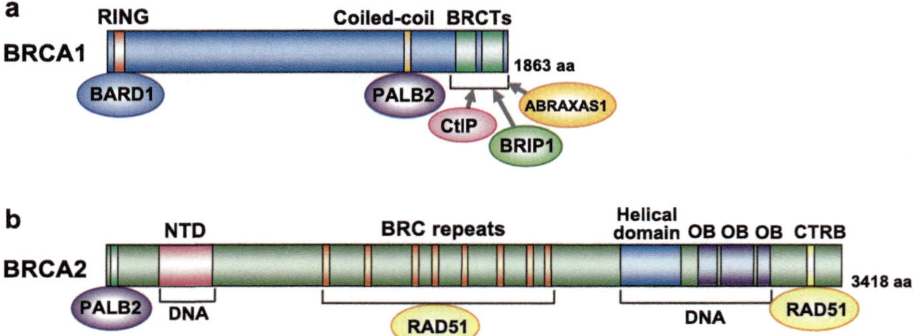

Fig. 1 Structure of BRCA1 and BRCA2. (**a**) BRCA1 has a RING domain in the N-terminal region, and a coiled-coil domain and two BRCT domains in the C-terminal region. BRCA1 binds to BARD1 via the RING domain. The coiled-coil domain of *BRCA1* mediates complex formation with PALB2. The BRCT domains bind to BRIP1, CtIP, and ABRAXAS1. (**b**) BRCA2 has a PALB2 binding domain at the N-terminal region, eight BRC repeats in the middle, and a helical domain, three oligonucleotide/oligosaccharide binding (OB) folds, and a RAD51-binding domain, (C-terminal RAD51-binding; CTRB) at the C-terminal region. The middle portion, which includes eight BRC repeats, binds to RAD51. A helical domain and three OB folds in the C-terminal domain and the N-terminal DNA binding domain (NTD) are DNA binding domains

3 Functions of BRCA1/2

3.1 HR

There are two major pathways that repair DNA DSBs: HR and non-homologous end joining (NHEJ). The HR is a pathway for error-free repair of DSBs uses the sister chromatid as a recombination template during the S/G2 phase of the cell cycle. By contrast, NHEJ repairs DSBs throughout the cell cycle by direct joining; however, it is error-prone and frequently causes deletion or insertion mutations in DNA [36].

The choice of which pathway is used to repair DSB is determined by DNA end resection, which is the processing of DNA ends to generate 3′ single strands. BRCA1/BARD1 competes with 53BP1 and promote end resection to proceed to HR pathway. DNA endoresection is initiated by the nuclease MRN complex, which comprises endonuclease MRE11, RAD50, and NBS1, to create short ssDNA (Fig. 2a). CtIP promotes end resection by MRE11. BRCA1/BARD1 plays a role in DNA end resection by interacting with the MRN complex and CtIP. ssDNA is occupied by replication protein A (RPA), which is then replaced by RAD51. Replacement of RPA by RAD51 is mediated by BRCA1-PALB2-BRCA2, thereby forming a RAD51-ssDNA nucleoprotein filament. The RAD51-ssDNA nucleoprotein filament functions in the homology search by invading duplex DNA molecules and facilitating base-pairing with complementary DNA sequences. BRCA1/BARD1

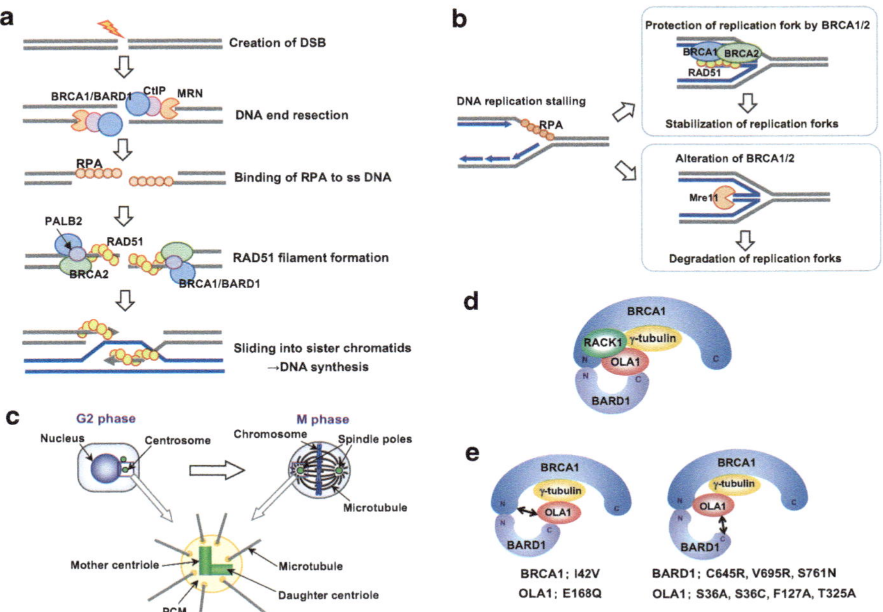

Fig. 2 Functions of BRCA1 and BRCA2. (**a**) The HR repair pathway. BRCA1/BARD1 promotes end resection in competition with 53BP1. DNA end resection is initiated by the nuclease MRN complex with CtIP. BRCA1/BARD1 is implicated in DNA end resection via interaction with the MRN complex and CtIP. ssDNA is covered by RPA, which is then replaced by RAD51; this process is mediated BRCA1-PALB2-BRCA2. The RAD51 filament functions in the homology search by invading duplex DNA. (**b**) DNA replication fork protection. When DNA replication forks are stalled, nascent DNA strands pair with each other and the fork regresses (fork reversal). BRCA1/2 prevents MRE11-mediated degradation of the free DNA end stabilizing RAD51 filaments at the stalled forks. Alterations in *BRCA1/2* restore MRE11-mediated degradation of free DNA ends, thereby collapsing the DNA replication fork. (**c**) Structure of the centrosome at G2 phase and the spindle poles at the mitotic phase. The centrosome comprises a pair of centrioles, mother centriole and daughter centriole, surrounded by pericentriolar material (PCM). (**d**) Model of the BRCA1/BARD1/OLA1/RACK1 complex. The N-terminal region of BRCA1 binds to the N-terminal region of BARD1. OLA1 binds to the N-terminal region of BRCA1, the C-terminal region of BARD1, and γ-tubulin. The middle portion of BRCA1 interacts with OLA1 via γ-tubulin. RACK1 binds to OLA1, the N-terminal region of BRCA1, and γ-tubulin. "N" indicates the N-terminal region. "C" indicates the C-terminal region. (**e**) Model of the conformational changes in the *BRCA1*/BARD1/OLA1/γ-tubulin complex induced by variants of BRCA1, BARD1, or OLA1. The I42V variant in BRCA1 and the E168Q variant in OLA1 impair binding of OLA1 to the N-terminal region of BRCA1. The C645R, V695L, and S761N variants in BARD1 and the S36A, S36C, F127A, and T325A variants in OLA1 impair binding of OLA1 to the C-terminal region of BARD1. These variants cause centrosome amplification. "N" indicates the N-terminal region. "C" indicates the C-terminal region

contributes to this homologous pairing. Therefore, BRCA1/BARD1 is involved in multiple steps of the HR repair pathway [4, 37, 38].

3.2 Stabilization of Replication Forks

When the DNA replication machinery encounters DNA lesions, or nucleotides become a limiting factor, DNA replication stalls. Newly synthesized DNA strands pair with each other, and the fork regresses (fork reversal), resulting a four-armed DNA structure that has free DNA end (Fig. 2b). BRCA1/2 prevents degradation of this free DNA end, which is mediated by the MRE11 nuclease, by stabilizing RAD51 filaments at the stalled forks. Therefore, BRCA1/2 prevent DNA damage. Alteration of BRCA1/2 results in the failure to protect replication forks, leading to collapse [37, 38].

3.3 Prevention of R-Loop Accumulation

R-loops, which comprise an RNA-DNA hybrid with a displaced ssDNA, occur on sites at which strong DNA secondary structures are formed due to perturbation of transcription or transcription-coupled processes such as mRNA splicing. Since an R-loop stalls the DNA replication machinery, cells avoid R-loop accumulation by preventing or removing them. BRCA1/BARD1 and BRCA2 prevent R-loop accumulation [37, 39].

3.4 Centrosome Regulation

Centrosomes regulate cell shape, polarity, and motility, in addition to formation of the mitotic spindle [14, 40]. The centrosome comprises a pair of centrioles, the mother and daughter centrioles, which are surrounded by the pericentriolar material (PCM) (Fig. 2c). Centrosomes are duplicated once per cell cycle, a process that is precisely controlled [41].

BRCA1 and BARD1, which localize to the centrosome throughout the cell cycle [42], function during centriole duplication [42–44]. BRCA1 interacts with major components of centrosomes: γ-tubulin [45] and mitotic kinases Aurora A and Polo-like kinase1 (PLK1) [46, 47]. BRCA1/BARD1 ubiquitinates centrosome proteins, including γ-tubulin [43]. We identified Obg-like ATPase 1 (OLA1) as a BARD1-interacting protein [48] and the receptor for activated C kinase 1 (RACK1) as an OLA1-interacting protein [49] (Fig. 2d). Aberrant expression of OLA1 and RACK1 occurs in many malignancies [50–54]. BRCA1 binds directly to OLA1 and RACK1, and functions to regulate centriole duplication together with these proteins. RACK1 regulates centriole duplication by controlling the centrosomal localization of BRCA1, as well as PLK1 phosphorylation by Aurora A [19, 20, 49].

A number of BRCA1 variants derived from familial breast cancers cause centrosome amplification in breast cancer cells [55]. Interestingly, some BRCA1 variants fail to regulate centrosome number; however, they are proficient in HR activity [8]. Cancer-derived variants of BRCA1, BARD1, OLA1, and RACK1 abolish their bindings each other. These variants and aberrant expression of these proteins cause centrosome amplification due to centriole overduplication only in mammary tissue-derived cells (Fig. 2e) [18, 19, 48, 49, 56]. These findings suggest that the BRCA1, together with these interacting proteins, regulates centrosomal numbers and contributes to tumor suppression.

Interestingly, the number of centrioles in cells with two γ-tubulin spots is higher in mammary tissue-derived cells than in cells derived from other tissues, suggesting that the efficiency of centriole duplication might be higher than that in cells derived from other tissues [49]. Thus, mammary cells might be sensitive to abnormality of centrosome proteins such as BRCA1-associated proteins, resulting in tissue-specific carcinogenesis.

BRCA2 also localizes to centrosomes and is involved in centriole duplication [57, 58]. Furthermore, BRCA2 interacts with a cytoskeletal cross-linker protein, plectin, and controls the position of the centrosome [59].

3.5 Other Functions

BRCA1/2 functions in multiple cellular processes in addition to the functions described above. The N-terminal region of BRCA1 has E3 ubiquitin ligase activity, which is enhanced significantly by forming a heterodimer with BARD1 [60]. The C-terminal region of BRCA1 interacts with RNA helicase A, a component of RNA polymerase II (Pol II) [61], and activates transcription. Since pathogenic variants abolish BRCA1-medicated transcriptional activation, functional assays were used to analyze a number of BRCA1 C-terminal missense variants to predict pathogenicity [6, 62, 63]. BRCA1 binds to several transcription factors, including p53, c-Myc, and GATA3. Furthermore, BRCA1 is involved in DNA damage checkpoint control, regulation of estrogen receptor α (ERα), apoptosis, and differentiation of luminal progenitor in breast tissues [6, 35, 62].

BRCA2 also regulates DNA damage checkpoint control [35]. In addition, BRCA2 is phosphorylated by PLK1, and regulates cytokinesis, a critical final step of cell division [64].

4 Evaluation of HR Activity

4.1 Evaluation of HR Activity in HBOC

Genetic alterations in HR factors in addition to *BRCA1/2* cause hereditary cancers such as HBOC [35]. Therefore, evaluation of HR activity in cells is important to estimate the pathogenicity of HR factor variants. As described above, the DR-GFP assay has been used widely for this purpose. In addition, Ikegami et al. developed a functional assay to evaluate the HR activity of BRCA2 variants by assessing sensitivity of *BRCA2*-deficient cells to PARP inhibitor [65].

4.2 Novel Assay to Evaluate HR Activity in Cells

To evaluate HR activity easily in cells, we developed the Assay for Site-specific HR Activity (ASHRA) (Fig. 3) [33]. In ASHRA, an expression vector for gRNA and Cas9 and a donor vector were co-transfected into the cells. The Cas9 endonuclease introduces a DSB specifically into the endogenous target locus of the genome. For our analysis, we chose the β-actin gene (*ACTB*), which is stably transcribed in cells, as a target gene. The donor vector contains a marker sequence flanked by two arms homologous to the target locus as a template for HR. Two days after transfection, genomic DNA was extracted. When the DSB is repaired by HR, the marker sequence is knocked-in to the target locus. HR activity is evaluated by quantifying the knock-in frequency by quantitative PCR. We confirmed that knockdown of BRCA1 or RAD51, but not that of non-HR factors, reduces the knock-in frequency [33].

4.3 A Novel HR Assay to Detect Intermediate HR Activity

Using ASHRA, we examined the HR activity of 30 BRCA1 missense variants at the N-terminal region that were previously analyzed using the DR-GFP assay [8, 66]. The DR-GFP assay categorized HR activity only as HR-proficient or HR-deficient, whereas ASHRA identified 10 *BRCA1* variants as having intermediate HR activity, which were not distinguished by the DR-GFP assay [34]. Interestingly, the HR activity of these BRCA1 variants, as assessed by ASHRA, correlated significantly with the survival rates of cells expressing BRCA1 variants after exposure to the PARP inhibitor olaparib.

The BRCA1-R1699Q variant moderately elevates cancer risk [67, 68]. The BRCA1-V1736A variant increases the risk of ovarian cancer through biallelic variation [69]. To investigate the significance of intermediate HR activity determined by ASHRA, we analyzed the HR activity of these BRCA1 variants. ASHRA detected the intermediate HR activity of BRCA1-R1699Q and -V1736A; in addition, these

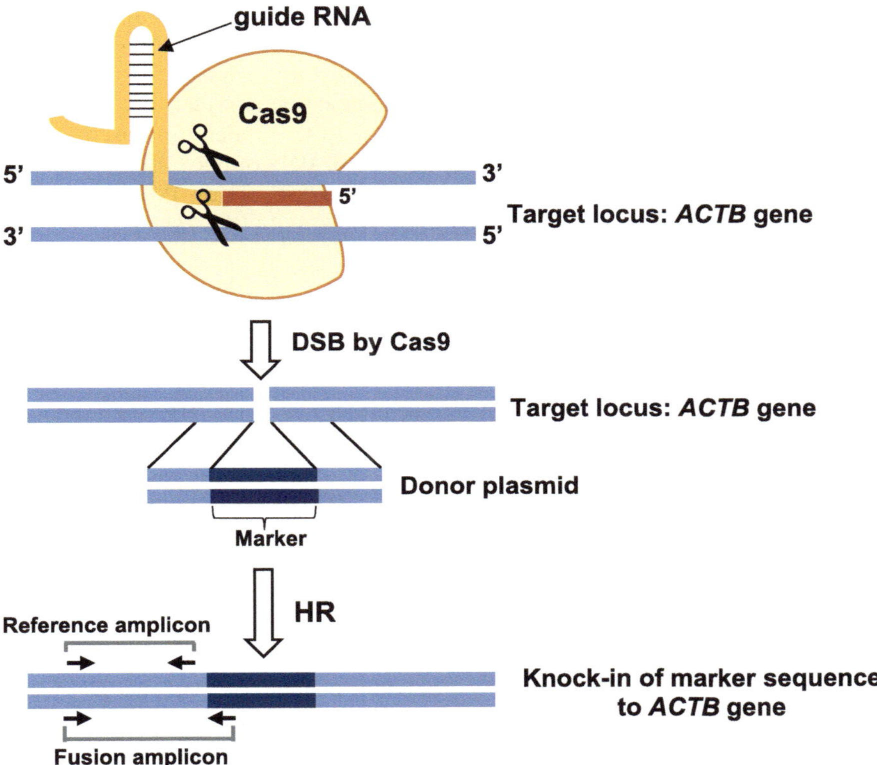

Fig. 3 Schematic of ASHRA. The Cas9 endonuclease creates double-strand breaks (DSBs) at the target site in the genome. When the DSBs are repaired by HR using a donor plasmid containing a marker sequence flanked by two arms homologous to the target site as a template, the marker sequence is knocked-in to the target site. HR activity in cells is evaluated by measuring the knock-in frequency of the marker sequence by quantitative PCR

variants showed intermediate sensitivity to olaparib. The DR-GFP assay categorized these variants as HR-deficient [9, 10]. These results suggest that HR activity determined by ASHRA can predict cancer risk and sensitivity to PARP inhibitors more accurately than the conventional DR-GFP assay.

4.4 *Role of HR Activity in Cancer Treatment*

Various sporadic cancers are HR-deficient [70]. The characteristics of tumors with germline *BRCA1/2* mutations are referred to as BRCAness, which is observed in some sporadic tumors with somatic mutations or methylation of *BRCA1/2*, or inactivation of HR factors other than *BRCA1/2* [71]. More than half of high-grade serous ovarian cancers are HR-deficient. Alterations of *BRCA1/2* are found in more than half of HR-deficient high-grade serous ovarian cancers, whereas alterations in other HR factors account for the remainder [72].

Genetic testing for HR factors, as well as genomic scar assays, is used clinically to detect HR deficiency and to stratify patients in order to identify those likely to benefit from PARP inhibitors. However, a number of VUS of HR factors have been identified, and genomic scar assays do not detect resistance to PARP inhibitors mediated by revertant mutations, which restore HR activity [73–75].

4.5 Mechanisms Underlying Resistance to PARP Inhibitors

Various mechanisms underlie primary and acquired resistance to PARP inhibitors [30–32]. Two major mechanisms are restoration of HR activity and protection of replication forks. HR can be restored by acquired mutations in HR genes or via increased activity of effector proteins that mediate HR activity. Acquired mutations restore the reading frame and allow expression of the entire protein. Tumors in which the *BRCA1* gene is silenced via promoter hypermethylation become resistant due to loss of hypermethylation and re-expression of *BRCA1*. Furthermore, restoration of HR activity is induced by suppression of the NHEJ pathway. 53BP1 regulates pathway choices, HR or NHEJ, in competition with BRCA1. Since 53BP1 collaborates with RIF1, REV7, and the Shieldin complex to prevents DNA end resection, these alterations trigger resistance by downregulating the NHEJ pathway. Protection of replication forks is caused by reduced recruitment of MRE11 and another DNA endonuclease, MUS81, leading to resistance to PARP inhibitors.

In addition, alterations in the PARP or PAR glycohydrolase proteins cause resistance to PARP inhibitors [31]. Similar to other anti-cancer agents, resistance is caused by increased drug efflux and reduced drug influx. Furthermore, Schlafen 11 (SFLN11) binds to replication forks in response to replication stress, thereby blocking further replication. SFLN11 sensitizes cells to a broad range of anti-cancer agents, including PARP inhibitors and platinum agents. Lack of SFLN11 expression in particular is involved in resistance to DNA-targeting anti-cancer agents [76, 77].

4.6 A Novel Mechanism of Resistance to PARP Inhibitors

Among the 30 BRCA1 missense variants analyzed, only the C61G variant shows significant discordance between HR activity and sensitivity to olaparib in HeLa cells, but not in MCF7 cells [34]. Interestingly, a similar C64G variant did not show these phenotypes. The C61G and C64G variants occur at the zinc-binding residues of the RING domain, resulting in defects in binding to BARD1, E3 ubiquitin ligase activity, and HR activity [8, 78, 79].

A literature search to identify different phenotypes of the C61G and C64G variants revealed that the BRCA1-C61G variant (but not C64G variant) functions to coactivate transcription by activating transcription factor 1 (ATF1) (similar to wild-type BRCA1) [80]. ATF1, which belongs to the c-AMP response element-binding protein/activating transcription factor (CREB/ATF) family, activates gene transcription to regulate cell proliferation and survival [81]. We found that BRCA1-C61G (but not the C64G variant) binds to ATF1 and activates transcription of *NRAS* and *BIRC2*, which are involved in cell proliferation and survival, respectively, similar to wild-type BRCA1 [34] (Fig. 4a). ATF1 was expressed at markedly higher levels in HeLa cells than in MCF7 cells. Furthermore, we found MCF10A is another ATF1-low cell line, and BT-549 is another ATF1-high cell line. Exogenous expression of

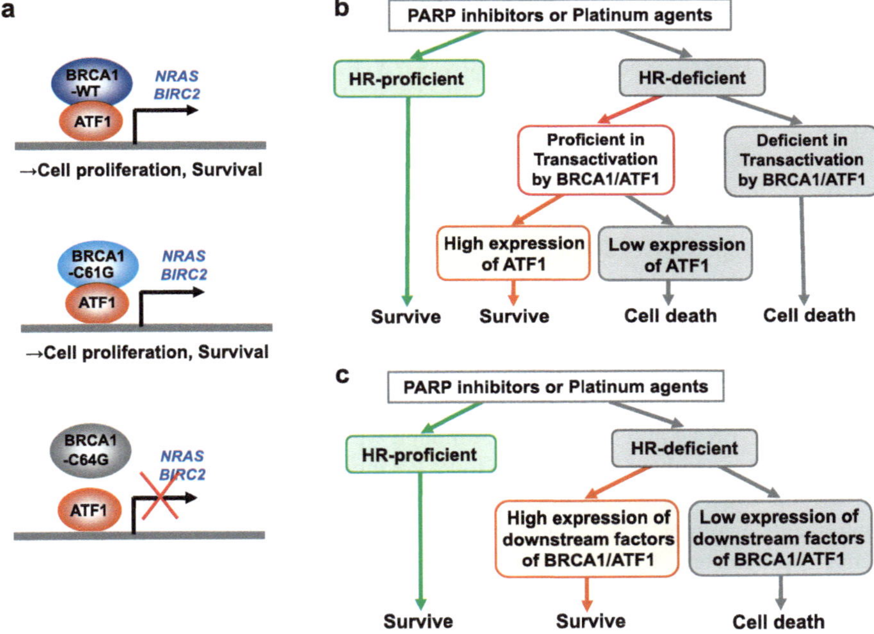

Fig. 4 Schematic illustrating ATF1-dependent sensitivity to PARP inhibitors and cisplatin. (**a**) Wild-type BRCA1 and the BRCA1-C61G variant bind to ATF1 and activates transcription of *NRAS* and *BIRC2*. By contrast, BRCA1-C64G fails binds to ATF1 and so does not activate the transcription of *NRAS* and *BIRC2*. (**b**) When treated with PARP inhibitors or platinum agents, HR-proficient cells repair DNA and survive. HR-deficient cells cannot repair DNA damage by HR. However, in cells proficient in BRCA1/ATF1-mediated transcription, cell survival is dependent on the expression level of ATF1. High ATF1-expressing cells proficient in BRCA1/ATF1-mediated transcription survive, but low-expressing cells die. HR-deficient cells harboring BRCA1-C61G or altered non-BRCA1 HR factors such as BRCA2, but possessing wild-type BRCA1, activate ATF1-mediated transcription and survive. Therefore, ATF1 might be a good biomarker for drug resistance in tumor cells. (**c**) In HR-deficient tumors, genes downstream of BRCA1/ATF1 transactivation might be good biomarkers for sensitivity to PARP inhibitors and platinum agents, regardless of BRCA1/ATF1 transactivation

ATF1 caused olaparib resistance in BRCA1-C61G–expressing ATF1-low cells. By contrast, knockdown of ATF1 increased the sensitivity of BRCA1-C61G–expressing ATF1-high cells to olaparib. The level of ATF1 expression did not affect HR activity. These data suggest that high expression of the ATF1 protein confers resistance to olaparib in cells expressing BRCA1-C61G, independent of HR activity (Fig. 4b).

4.7 BRCA1/ATF1-Mediated Transactivation Confers Resistance to Cisplatin

BRCA1-C61G is HR-deficient, but functions as a coactivator of ATF1-regulated transcription, which causes resistance to olaparib in the presence of high ATF1 expression. Thus, we speculated that BRCA1/ATF1-mediated transcription confers resistance to olaparib in HR-deficient cells due to alterations in other HR factors. As expected, we found that ATF1 overexpression induced olaparib resistance in BRCA2- or RAD51-knockdown MCF7 cells, but not in BRCA1-knockdown cells. Similar results were obtained for cells treated with cisplatin [34]. These data suggest that BRCA1/ATF1-mediated transcription induces the resistance to olaparib and cisplatin upon knockdown of BRCA2 or RAD51. Therefore, the level of ATF1 expression could be a biomarker for the efficacy of PARP inhibitors and platinum agents against tumors that are HR-deficient, but proficient in BRCA1/ATF1-mediated transcription (e.g., *BRCA2*-deficient tumors) (Fig. 4b).

Tian et al. identified ATF1-target genes [81]. We found that there are two types of ATF1-target genes, BRCA1-dependent (*NRAS* and *BIRC2*) and BRCA1-independent (*BRAF* and *MYC*) [34]. Although we did not examine whether NRAS and BIRC2 are involved in resistance to PARP inhibitors and platinum agents, their expression levels might be a biomarker for the efficacy of PARP inhibitors and platinum agents against HR-deficient cells (Fig. 4c). However, it is possible that other genes upregulated by BRCA1/ATF1 might be responsible for resistance. Identification of the responsible genes might make it possible to develop therapies that overcome resistance to PARP inhibitors and platinum agents.

BRCA1-C61G is an important founder mutation in the Polish population; therefore, it is analyzed in panel tests for HBOC diagnosis and for cancer treatment [82, 83]. Therefore, resistance to PARP inhibitors and platinum agents induced by BRCA1/ATF1-mediated transcription might be considered in the Polish population.

Overexpression of ATF1, NRAS, and BIRC2 occurs in some malignancies, including breast and ovarian cancers (COSMIC, the Catalogue Of Somatic Mutations In Cancer (https://cancer.sanger.ac.uk/cosmic)). *ATF1* forms a fusion gene with *EWSR1* or *FUS* in sarcomas such as clear cell sarcoma and angiomatoid fibrous histiocytoma [84, 85]. Therefore, ATF1-fusion gene products might be involved in resistance of these sarcomas to chemotherapy.

5 Conclusions

HR activity is important for the tumor suppressor functions of BRCA1/2. We used the CRISPR/Cas9 system to develop ASHRA, an assay designed to evaluate HR activity. ASHRA can measure HR activity quantitatively; the results show that HR activity correlates with cancer risk and sensitivity to PARP inhibitors. Furthermore, we identified a novel mechanism involving BRCA1/ATF1-mediated transcription that underlies resistance to olaparib and cisplatin. This assay will contribute to classification of VUS of HR factors that might be involved in cancer risk and sensitivity to PARP inhibitors and platinum agents. It will also be useful for identifying novel resistance mechanisms and target molecules to facilitate development of effective cancer therapies.

BRCA1/2 plays role in multiple cellular processes. However, the role of BRCA1/2 alterations in mechanisms of underlying tissue-specific carcinogenesis remains unclear. The roles played by BRCA1/2 alterations in cancer risk and sensitivity to anti-cancer agents seem to be different. Thus, functional assays to evaluate the effects of *BRCA1/2* variants should be changed dependent on the purpose of the analysis: prediction of cancer risk or sensitivity to anti-cancer agents.

References

1. Miki Y, Swensen J, Shattuck-Eidens D, Futreal PA, Harshman K, Tavtigian S, et al. A strong candidate for the breast and ovarian cancer susceptibility gene BRCA1. Science. 1994;266(5182):66–71.
2. Wooster R, Bignell G, Lancaster J, Swift S, Seal S, Mangion J, et al. Identification of the breast cancer susceptibility gene BRCA2. Nature. 1995;378(6559):789–92.
3. Momozawa Y, Sasai R, Usui Y, Shiraishi K, Iwasaki Y, Taniyama Y, et al. Expansion of cancer risk profile for BRCA1 and BRCA2 pathogenic variants. JAMA. Oncologia. 2022;8:871.
4. Zhao W, Wiese C, Kwon Y, Hromas R, Sung P. The BRCA tumor suppressor network in chromosome damage repair by homologous recombination. Annu Rev Biochem. 2019;88:221–45.
5. Gorodetska I, Kozeretska I, Dubrovska A. BRCA genes: the role in genome stability, cancer Stemness and therapy resistance. J Cancer. 2019;10(9):2109–27.
6. Takaoka M, Miki Y. BRCA1 gene: function and deficiency. Int J Clin Oncol. 2018;23(1):36–44.
7. Pierce AJ, Hu P, Han M, Ellis N, Jasin M. Ku DNA end-binding protein modulates homologous repair of double-strand breaks in mammalian cells. Genes Dev. 2001;15(24):3237–42.
8. Ransburgh DJ, Chiba N, Ishioka C, Toland AE, Parvin JD. Identification of breast tumor mutations in BRCA1 that abolish its function in homologous DNA recombination. Cancer Res. 2010;70(3):988–95.
9. Anantha RW, Simhadri S, Foo TK, Miao S, Liu J, Shen Z, et al. Functional and mutational landscapes of BRCA1 for homology-directed repair and therapy resistance. elife. 2017;6:6.
10. Bouwman P, van der Gulden H, van der Heijden I, Drost R, Klijn CN, Prasetyanti P, et al. A high-throughput functional complementation assay for classification of BRCA1 missense variants. Cancer Discov. 2013;3(10):1142–55.
11. Guidugli L, Carreira A, Caputo SM, Ehlen A, Galli A, Monteiro AN, et al. Functional assays for analysis of variants of uncertain significance in BRCA2. Hum Mutat. 2014;35(2):151–64.

12. Guidugli L, Pankratz VS, Singh N, Thompson J, Erding CA, Engel C, et al. A classification model for BRCA2 DNA binding domain missense variants based on homology-directed repair activity. Cancer Res. 2013;73(1):265–75.
13. Guidugli L, Shimelis H, Masica DL, Pankratz VS, Lipton GB, Singh N, et al. Assessment of the clinical relevance of BRCA2 missense variants by functional and computational approaches. Am J Hum Genet. 2018;102(2):233–48.
14. Nigg EA, Holland AJ. Once and only once: mechanisms of centriole duplication and their deregulation in disease. Nat Rev Mol Cell Biol. 2018;19(5):297–312.
15. Nigg EA, Schnerch D, Ganier O. Impact of centrosome aberrations on chromosome segregation and tissue architecture in cancer. Cold Spring Harb Symp Quant Biol. 2017;82:137–44.
16. LoMastro GM, Holland AJ. The emerging link between centrosome aberrations and metastasis. Dev Cell. 2019;49(3):325–31.
17. Gonczy P. Centrosomes and cancer: revisiting a long-standing relationship. Nat Rev Cancer. 2015;15(11):639–52.
18. Otsuka K, Yoshino Y, Qi H, Chiba N. The function of BARD1 in centrosome regulation in cooperation with BRCA1/OLA1/RACK1. Genes (Basel). 2020;11(8):842.
19. Yoshino Y, Fang Z, Qi H, Kobayashi A, Chiba N. Dysregulation of the centrosome induced by BRCA1 deficiency contributes to tissue-specific carcinogenesis. Cancer Sci. 2021;112:1679.
20. Yoshino Y, Chiba N. Roles of RACK1 in centrosome regulation and carcinogenesis. Cell Signal. 2022;90:110207.
21. Ray Chaudhuri A, Nussenzweig A. The multifaceted roles of PARP1 in DNA repair and chromatin remodelling. Nat Rev Mol Cell Biol. 2017;18(10):610–21.
22. Pommier Y, O'Connor MJ, de Bono J. Laying a trap to kill cancer cells: PARP inhibitors and their mechanisms of action. Sci Transl Med. 2016;8(362):362ps17.
23. Ray Chaudhuri A, Hashimoto Y, Herrador R, Neelsen KJ, Fachinetti D, Bermejo R, et al. Topoisomerase I poisoning results in PARP-mediated replication fork reversal. Nat Struct Mol Biol. 2012;19(4):417–23.
24. Maya-Mendoza A, Moudry P, Merchut-Maya JM, Lee M, Strauss R, Bartek J. High speed of fork progression induces DNA replication stress and genomic instability. Nature. 2018;559(7713):279–84.
25. Rottenberg S, Jaspers JE, Kersbergen A, van der Burg E, Nygren AO, Zander SA, et al. High sensitivity of BRCA1-deficient mammary tumors to the PARP inhibitor AZD2281 alone and in combination with platinum drugs. Proc Natl Acad Sci U S A. 2008;105(44):17079–84.
26. Farmer H, McCabe N, Lord CJ, Tutt AN, Johnson DA, Richardson TB, et al. Targeting the DNA repair defect in BRCA mutant cells as a therapeutic strategy. Nature. 2005;434(7035):917–21.
27. Bryant HE, Schultz N, Thomas HD, Parker KM, Flower D, Lopez E, et al. Specific killing of BRCA2-deficient tumours with inhibitors of poly(ADP-ribose) polymerase. Nature. 2005;434(7035):913–7.
28. McCabe N, Turner NC, Lord CJ, Kluzek K, Bialkowska A, Swift S, et al. Deficiency in the repair of DNA damage by homologous recombination and sensitivity to poly(ADP-ribose) polymerase inhibition. Cancer Res. 2006;66(16):8109–15.
29. Pilie PG, Gay CM, Byers LA, O'Connor MJ, Yap TA. PARP inhibitors: extending benefit beyond BRCA-mutant cancers. Clin Cancer Res. 2019;25(13):3759–71.
30. D'Andrea AD. Mechanisms of PARP inhibitor sensitivity and resistance. DNA Repair (Amst). 2018;71:172–6.
31. Zheng F, Zhang Y, Chen S, Weng X, Rao Y, Fang H. Mechanism and current progress of poly ADP-ribose polymerase (PARP) inhibitors in the treatment of ovarian cancer. Biomed Pharmacother. 2020;123:109661.
32. Reilly NM, Novara L, Di Nicolantonio F, Bardelli A. Exploiting DNA repair defects in colorectal cancer. Mol Oncol. 2019;13(4):681–700.
33. Yoshino Y, Endo S, Chen Z, Qi H, Watanabe G, Chiba N. Evaluation of site-specific homologous recombination activity of BRCA1 by direct quantitation of gene editing efficiency. Sci Rep. 2019;9(1):1644.

34. Endo S, Yoshino Y, Shirota M, Watanabe G, Chiba N. BRCA1/ATF1-mediated transactivation is involved in resistance to PARP inhibitors and cisplatin. Cancer Research Communications. 2021;1(2):90–105.
35. Nielsen FC, van Overeem HT, Sorensen CS. Hereditary breast and ovarian cancer: new genes in confined pathways. Nat Rev Cancer. 2016;16(9):599–612.
36. Scully R, Panday A, Elango R, Willis NA. DNA double-strand break repair-pathway choice in somatic mammalian cells. Nat Rev Mol Cell Biol. 2019;20(11):698–714.
37. Tarsounas M, Sung P. The antitumorigenic roles of BRCA1-BARD1 in DNA repair and replication. Nat Rev Mol Cell Biol. 2020;21(5):284–99.
38. Chen CC, Feng W, Lim PX, Kass EM, Jasin M. Homology-directed repair and the role of BRCA1, BRCA2, and related proteins in genome integrity and cancer. Annu Rev Cancer Biol. 2018;2:313–36.
39. Santos-Pereira JM, Aguilera A. R loops: new modulators of genome dynamics and function. Nat Rev Genet. 2015;16(10):583–97.
40. Conduit PT, Wainman A, Raff JW. Centrosome function and assembly in animal cells. Nat Rev Mol Cell Biol. 2015;16(10):611–24.
41. Fujita H, Yoshino Y, Chiba N. Regulation of the centrosome cycle. Mol Cell Oncol. 2016;3(2):e1075643.
42. Sankaran S, Starita LM, Groen AC, Ko MJ, Parvin JD. Centrosomal microtubule nucleation activity is inhibited by BRCA1-dependent ubiquitination. Mol Cell Biol. 2005;25(19):8656–68.
43. Starita LM, Machida Y, Sankaran S, Elias JE, Griffin K, Schlegel BP, et al. BRCA1-dependent ubiquitination of gamma-tubulin regulates centrosome number. Mol Cell Biol. 2004;24(19):8457–66.
44. Ko MJ, Murata K, Hwang DS, Parvin JD. Inhibition of BRCA1 in breast cell lines causes the centrosome duplication cycle to be disconnected from the cell cycle. Oncogene. 2006;25(2):298–303.
45. Hsu LC, Doan TP, White RL. Identification of a gamma-tubulin-binding domain in BRCA1. Cancer Res. 2001;61(21):7713–8.
46. Ertych N, Stolz A, Stenzinger A, Weichert W, Kaulfuss S, Burfeind P, et al. Increased microtubule assembly rates influence chromosomal instability in colorectal cancer cells. Nat Cell Biol. 2014;16(8):779–91.
47. Zou J, Rezvani K, Wang H, Lee KS, Zhang D. BRCA1 downregulates the kinase activity of polo-like kinase 1 in response to replication stress. Cell Cycle. 2013;12(14):2255–65.
48. Matsuzawa A, Kanno S, Nakayama M, Mochiduki H, Wei L, Shimaoka T, et al. The BRCA1/BARD1-interacting protein OLA1 functions in centrosome regulation. Mol Cell. 2014;53(1):101–14.
49. Yoshino Y, Qi H, Kanazawa R, Sugamata M, Suzuki K, Kobayashi A, et al. RACK1 regulates centriole duplication by controlling localization of BRCA1 to the centrosome in mammary tissue-derived cells. Oncogene. 2019;38(16):3077–92.
50. Sun H, Luo X, Montalbano J, Jin W, Shi J, Sheikh MS, et al. DOC45, a novel DNA damage-regulated nucleocytoplasmic ATPase that is overexpressed in multiple human malignancies. Mol Cancer Res. 2010;8(1):57–66.
51. Duff D, Long A. Roles for RACK1 in cancer cell migration and invasion. Cell Signal. 2017;35:250–5.
52. Cao XX, Xu JD, Liu XL, Xu JW, Wang WJ, Li QQ, et al. RACK1: a superior independent predictor for poor clinical outcome in breast cancer. Int J Cancer. 2010;127(5):1172–9.
53. Cao XX, Xu JD, Xu JW, Liu XL, Cheng YY, Wang WJ, et al. RACK1 promotes breast carcinoma proliferation and invasion/metastasis in vitro and in vivo. Breast Cancer Res Treat. 2010;123(2):375–86.
54. Al-Reefy S, Mokbel K. The role of RACK1 as an independent prognostic indicator in human breast cancer. Breast Cancer Res Treat. 2010;123(3):911; author reply 2
55. Kais Z, Chiba N, Ishioka C, Parvin JD. Functional differences among BRCA1 missense mutations in the control of centrosome duplication. Oncogene. 2012;31(6):799–804.

56. Yoshino Y, Qi H, Fujita H, Shirota M, Abe S, Komiyama Y, et al. BRCA1-interacting protein OLA1 requires interaction with BARD1 to regulate centrosome number. Mol Cancer Res. 2018;16:1499–511.
57. Nakanishi A, Han X, Saito H, Taguchi K, Ohta Y, Imajoh-Ohmi S, et al. Interference with BRCA2, which localizes to the centrosome during S and early M phase, leads to abnormal nuclear division. Biochem Biophys Res Commun. 2007;355(1):34–40.
58. Han X, Saito H, Miki Y, Nakanishi A. A CRM1-mediated nuclear export signal governs cytoplasmic localization of BRCA2 and is essential for centrosomal localization of BRCA2. Oncogene. 2008;27(21):2969–77.
59. Niwa T, Saito H, Imajoh-ohmi S, Kaminishi M, Seto Y, Miki Y, et al. BRCA2 interacts with the cytoskeletal linker protein plectin to form a complex controlling centrosome localization. Cancer Sci. 2009;100(11):2115–25.
60. Hashizume R, Fukuda M, Maeda I, Nishikawa H, Oyake D, Yabuki Y, et al. The RING heterodimer BRCA1-BARD1 is a ubiquitin ligase inactivated by a breast cancer-derived mutation. J Biol Chem. 2001;276(18):14537–40.
61. Anderson SF, Schlegel BP, Nakajima T, Wolpin ES, Parvin JD. BRCA1 protein is linked to the RNA polymerase II holoenzyme complex via RNA helicase A. Nat Genet. 1998;19(3):254–6.
62. Zhang X, Li R. BRCA1-dependent transcriptional regulation: implication in tissue-specific tumor suppression. Cancers (Basel). 2018;10(12):513.
63. Fernandes VC, Golubeva VA, Di Pietro G, Shields C, Amankwah K, Nepomuceno TC, et al. Impact of amino acid substitutions at secondary structures in the BRCT domains of the tumor suppressor BRCA1: implications for clinical annotation. J Biol Chem. 2019;294(15):5980–92.
64. Takaoka M, Saito H, Takenaka K, Miki Y, Nakanishi A. BRCA2 phosphorylated by PLK1 moves to the midbody to regulate cytokinesis mediated by nonmuscle myosin IIC. Cancer Res. 2014;74(5):1518–28.
65. Ikegami M, Kohsaka S, Ueno T, Momozawa Y, Inoue S, Tamura K, et al. High-throughput functional evaluation of BRCA2 variants of unknown significance. Nat Commun. 2020;11(1):2573.
66. Towler WI, Zhang J, Ransburgh DJ, Toland AE, Ishioka C, Chiba N, et al. Analysis of BRCA1 variants in double-strand break repair by homologous recombination and single-strand annealing. Hum Mutat. 2013;34(3):439–45.
67. Moghadasi S, Meeks HD, Vreeswijk MP, Janssen LA, Borg A, Ehrencrona H, et al. The BRCA1 c. 5096G>A p.Arg1699Gln (R1699Q) intermediate risk variant: breast and ovarian cancer risk estimation and recommendations for clinical management from the ENIGMA consortium. J Med Genet. 2018;55(1):15–20.
68. Spurdle AB, Whiley PJ, Thompson B, Feng B, Healey S, Brown MA, et al. BRCA1 R1699Q variant displaying ambiguous functional abrogation confers intermediate breast and ovarian cancer risk. J Med Genet. 2012;49(8):525–32.
69. Domchek SM, Tang J, Stopfer J, Lilli DR, Hamel N, Tischkowitz M, et al. Biallelic deleterious BRCA1 mutations in a woman with early-onset ovarian cancer. Cancer Discov. 2013;3(4):399–405.
70. Riaz N, Blecua P, Lim RS, Shen R, Higginson DS, Weinhold N, et al. Pan-cancer analysis of bi-allelic alterations in homologous recombination DNA repair genes. Nat Commun. 2017;8(1):857.
71. Turner N, Tutt A, Ashworth A. Hallmarks of 'BRCAness' in sporadic cancers. Nat Rev Cancer. 2004;4(10):814–9.
72. Cancer Genome Atlas Research N. Integrated genomic analyses of ovarian carcinoma. Nature. 2011;474(7353):609–15.
73. Barber LJ, Sandhu S, Chen L, Campbell J, Kozarewa I, Fenwick K, et al. Secondary mutations in BRCA2 associated with clinical resistance to a PARP inhibitor. J Pathol. 2013;229(3):422–9.
74. Dhillon KK, Swisher EM, Taniguchi T. Secondary mutations of BRCA1/2 and drug resistance. Cancer Sci. 2011;102(4):663–9.
75. Kondrashova O, Nguyen M, Shield-Artin K, Tinker AV, Teng NNH, Harrell MI, et al. Secondary somatic mutations restoring RAD51C and RAD51D associated with acquired

resistance to the PARP inhibitor Rucaparib in high-grade ovarian carcinoma. Cancer Discov. 2017;7(9):984–98.
76. Murai J, Zhang H, Pongor L, Tang SW, Jo U, Moribe F, et al. Chromatin remodeling and immediate early gene activation by SLFN11 in response to replication stress. Cell Rep. 2020;30(12):4137–51 e6.
77. Murai J, Thomas A, Miettinen M, Pommier Y. Schlafen 11 (SLFN11), a restriction factor for replicative stress induced by DNA-targeting anti-cancer therapies. Pharmacol Ther. 2019;201:94–102.
78. Wu LC, Wang ZW, Tsan JT, Spillman MA, Phung A, Xu XL, et al. Identification of a RING protein that can interact in vivo with the BRCA1 gene product. Nat Genet. 1996;14(4):430–40.
79. Brzovic PS, Keeffe JR, Nishikawa H, Miyamoto K, Fox D 3rd, Fukuda M, et al. Binding and recognition in the assembly of an active BRCA1/BARD1 ubiquitin-ligase complex. Proc Natl Acad Sci U S A. 2003;100(10):5646–51.
80. Houvras Y, Benezra M, Zhang H, Manfredi JJ, Weber BL, Licht JD. BRCA1 physically and functionally interacts with ATF1. J Biol Chem. 2000;275(46):36230–7.
81. Tian J, Chang J, Gong J, Lou J, Fu M, Li J, et al. Systematic functional interrogation of genes in GWAS loci identified ATF1 as a key driver in colorectal cancer modulated by a promoter-enhancer interaction. Am J Hum Genet. 2019;105(1):29–47.
82. Szwiec M, Jakubowska A, Gorski B, Huzarski T, Tomiczek-Szwiec J, Gronwald J, et al. Recurrent mutations of BRCA1 and BRCA2 in Poland: an update. Clin Genet. 2015;87(3):288–92.
83. Menkiszak J, Chudecka-Glaz A, Gronwald J, Bedner R, Cymbaluk-Ploska A, Wezowska M, et al. Characteristics of selected clinical features in BRCA1 mutation carriers affected with breast cancer undergoing preventive female genital tract surgeries. Ginekol Pol. 2013;84(9):758–64.
84. Hocar O, Le Cesne A, Berissi S, Terrier P, Bonvalot S, Vanel D, et al. Clear cell sarcoma (malignant melanoma) of soft parts: a clinicopathologic study of 52 cases. Dermatol Res Pract. 2012;2012:984096.
85. Thway K, Fisher C. Angiomatoid fibrous histiocytoma: the current status of pathology and genetics. Arch Pathol Lab Med. 2015;139(5):674–82.

Part IV
Health and Labor Sciences Research Grant

Hereditary Breast and Ovarian Cancer and the National Health Insurance in Japan

Akira Hirasawa

Abstract Medical practices for hereditary breast and ovarian cancer (HBOC) are partially covered by public medical insurance in Japan. In 2020, *BRCA1* or *BRCA2* (*BRCA1/2*) genetic testing, surveillance, risk-reducing mastectomy (RRM), and risk-reducing salpingo-oophorectomy for *BRCA1/2* pathogenic variant carriers were approved for all patients with ovarian cancer and some patients with breast cancer under National Health Insurance. However, relatives of HBOC pedigrees and *BRCA1/2* pathogenic variant carriers who do not have breast or ovarian cancer are not covered by public health insurance. Furthermore, the clinical application of multigene panel testing (MGPT) is an urgent issue.

Keywords Hereditary breast and ovarian cancer (HBOC) · National health insurance · Cancer precision medicine · *BRCA1/2* genetic testing · Risk-reducing mastectomy · Risk-reducing salpingo-oophorectomy · Genetic counseling

1 Current State of Public Health Insurance Coverage for HBOC in Japan

Table 1 shows the National Health Insurance coverage for HBOC in Japan as of May 2023. HBOC medical treatment covered by health insurance is only for patients who have already developed breast or ovarian cancer. The covered treatments include the following: (1) guidance and genetic counseling costs for *BRCA1/2* genetic testing, (2) *BRCA1/2* genetic testing, (3) risk-reducing salpingo-oophorectomy (RRSO) and contralateral risk-reducing mastectomy (CRRM) for those diagnosed with HBOC among patients with breast cancer, (4) bilateral risk-reducing mastectomy (BRRM) for women who develop ovarian cancer and are

A. Hirasawa (✉)
Department of Clinical Genomic Medicine, Okayama University Graduate School of Medicine, Dentistry and Pharmaceutical Sciences, Okayama, Japan
e-mail: hir-aki45@okayama-u.ac.jp

D. Aoki et al. (eds.), *Practical Guide to Hereditary Breast and Ovarian Cancer*, https://doi.org/10.1007/978-981-99-5231-1_8

Table 1 Public health insurance coverage for hereditary breast cancer and ovarian cancer in Japan (as of May 2023)

	BRCA1/2 genetic testing		Surveillance				Risk-reducing surgery	
	HBOC diagnosis	Companion diagnostic	Breast *1	Ovary *2	Prostate*3	Pancreas*4	Risk-reducing mastectomy	Risk-reducing salpingo-oophorectomy
People without cancer	Not covered	–	Not covered	Not covered	Not covered	Not covered	Not covered	Not covered
Breast cancer patient	Covered	Covered	Covered	Covered	Not covered	Not covered	Covered	Not covered
Ovarian cancer patient	Covered	Covered	Covered	–	–	Not covered	Covered	–
Prostate cancer patient	Not covered	Covered	Not covered	–	–	Not covered	Not covered	–
Pancreatic cancer patient	Not covered	Covered	Not covered			Not covered	Not covered	Not covered
Patients with cancer other than breast cancer, ovarian cancer, prostate cancer, and pancreatic cancer	Not covered	Not covered	Not covered	Not covered	Not covered	Not covered	Not covered	Not covered

*1 Mammography, breast ultrasound, mammography MRI, etc.
*2 Transvaginal ultrasound, serum CA125 tumor marker, etc.
*3 Serum PSA
*4 MRI, magnetic resonance cholangiopancreatography (MRCP) or ultrasound endoscopy (sometimes replaced by contrast-enhanced CT), serum CA19–9, etc.
*5 Including patients with fallopian tube cancer or peritoneal cancer

diagnosed with HBOC, and (5) surveillance for patients diagnosed with HBOC. The facility regulations stipulated by the Ministry of Health, Labor, and Welfare in Japan must be satisfied when applying for insurance medical treatment for HBOC.

2 Definition of HBOC and Approach to Resolving Uninsured Matters

The narrow definition of HBOC is defined as "a cancer susceptibility syndrome, including breast and ovarian cancer, caused by germline pathogenic variants in *BRCA1/2*." Furthermore, a broader definition of HBOC includes the involvement of multiple genes related to susceptibility to breast or ovarian cancer beyond *BRCA1/2* [1]. In Japan, the "hereditary breast cancer and ovarian cancer syndromes" registered in 2020 are not cancer diseases based on this definition. HBOC is a disease with a pathogenic variant that increases susceptibility to related cancers, such as breast, ovarian, prostate, and pancreatic cancers, regardless of cancer development.

3 Issues That Japan Needs to Solve in Order to Aim for Seamless HBOC Medical Care

3.1 The Restriction on BRCA1/2 Genetic Testing for Confirming the Diagnosis of HBOC for Those Who Have Already Developed Breast or Ovarian Cancer Should Be Lifted

BRCA1/2 genetic testing for HBOC diagnosis is currently not covered by the public health insurance, except in patients with breast or ovarian cancer. Therefore, expanding insurance coverage for *BRCA1/2* genetic testing is necessary for the following reasons:

1. Individuals suspected of having an HBOC family history, regardless of whether they have breast or ovarian cancer.
2. Family members with known *BRCA1/2* pathogenic variant carriers in the family. First-degree relatives have a high pretest predictive probability of 50%, with 25% for second-degree relatives and 12.5% for third-degree relatives, possibly carrying the same variant as the proband).
3. Case of *BRCA1/2* pathogenic variant identified as presumed pathogenic germline variant (PGPV) by comprehensive genome profile (CGP) (Approximately 80% of the cases in which *BRCA1/2* pathogenic variants were identified in tumor tissues were derived from the germline [2, 3]).

3.2 Removal of Restrictions on Breast Surveillance Only for Those Who Have Already Developed Breast or Ovarian Cancer

For *BRCA1/2* pathogenic variant carriers, surveillance using mammography, breast ultrasonography, and mammographic MRI has been reported to improve the early detection rate and survival prognosis of breast cancer. However, breast cancer in *BRCA1/2* pathogenic variant carriers tends to develop at a young age (approximately 10 years), and women in their 30s and 40s often have dense mammary glands, making mammography less useful. The sensitivity of MRI is also known to be high. Although both those who have not yet developed breast or ovarian cancer and those who have already developed it are at a high risk of developing breast cancer, currently, public insurance only covers those who have already developed breast or ovarian cancer. Therefore, it is necessary to address the limitations of public insurance for patients who have already developed breast or ovarian cancers.

3.3 Canceling the Limitation of Ovarian Surveillance to Those Who Have Already Developed Breast Cancer

If RRSO is not selected in *BRCA1/2* pathogenic variant carriers, transvaginal ultrasonography and serum CA125 test are considered. However, ovarian cancer surveillance is currently only covered by health insurance for patients who have already developed breast cancer. Therefore, expanding insurance coverage for surveillance using transvaginal ultrasonography and tumor markers (CA125) is necessary for women with *BRCA1/2* pathogenic variants and should not be limited to those who have already developed breast cancer.

3.4 The Limitation of Risk-Reducing Salpingo-oophorectomy to Patients with Existing Breast Cancer Should Be Lifted

Risk-reducing salpingo-oophorectomy (RRSO) is performed only in women with HBOC and breast cancer. Medical law has also shown that the prognosis-improving effect of risk-reducing surgery in *BRCA1/2* pathogenic variant carriers is independent of the onset of breast or ovarian cancer [4]. Therefore, for women with HBOC, overcoming the limitations of RRSO for those who have already developed breast cancer will lead to a reduction in cancer deaths.

4 Risk-Reducing Mastectomy Should Be Lifted from Being Limited to Women with Pre-existing Breast or Ovarian Cancer

Among HBOCs, risk-reducing mastectomy (RRM) is covered by health insurance only for those who have already developed breast or ovarian cancer. However, it is required to lift these limitations for those who have already developed breast or ovarian cancer.

4.1 Remove the Restriction on Intractable Diseases for Remote Genetic Counseling Covered by Public Insurance

In the 2022 medical fee revision in Japan, remote genetic counseling was covered by public insurance but was limited to examinations for intractable diseases, and remote genetic counseling for hereditary tumors, including HBOC, was not permitted. As many hereditary tumors have autosomal dominant inheritance, genetic counseling involving the simultaneous participation of relatives may lead to cancer prevention for the entire family.

5 Toward Introducing Multigene Panel Testing (MGPT)

Advances in gene analysis technology using next-generation sequencing (NGS) have enabled genetic testing to be performed at low cost and with high efficiency. Since 2014, the widespread adoption of MGPT, which comprehensively examines many candidate genes, has replaced single-gene testing [5]. In the US, the Supreme Court did not recognize the patent on *BRCA1/2* acquired by Myriad Genetics in 2013, making it possible for other inspection companies to perform the analysis. However, in Japan, while *BRCA1/2* genetic testing has been included in public insurance coverage since 2020, the identification of hereditary tumor families besides HBOC has been omitted.

Our group performed MGPT in 230 epithelial ovarian cancer cases in Japan, of which 27 (11.8%) were *BRCA1/2*-positive. In addition, 14 patients (6.0%) carried germline variants of genes other than BRCA1/2 [6]. Although *BRCA1/2* genetic testing for all patients with epithelial ovarian cancer is covered by medical insurance under the 2020 medical fee revision in Japan, other genes that can be detected without MGPT may have been overlooked.

In the United States, in 2017, approximately 90% of tests performed on patients with breast and ovarian cancers were multigene panel tests, and full-scale introduction is an urgent issue in Japan as well [7]. The characteristics of the MGPT are summarized in Table 2.

Table 2 Characteristics of multigene panel testing

Simultaneous diagnosis of multiple hereditary tumor-related genes
Detects variants that are not identified by genetic testing for a single gene (mult-igene identification)
Enables detection of genes that cannot be inferred from the phenotype
Cheaper than examining a single gene multiple times and shorter TAT (turnaround time)
If a hereditary tumor is diagnosed, measures can be taken to reduce the risk or early detection in collaboration with related clinical departments
There is no medical justification for undue testing if no pathogenic variant is identified
It is possible to decide the treatment policy (drug sensitivity, surgical method, and irradiation)
Testing can be performed regardless of the presence or absence of cancer development

6 Conclusion

Cancer genomic medicine, which is a national policy of Japan, is defined as "medical care that optimizes treatment, predicts prognosis, and prevents the onset of cancer using the genomic information of tumors and normal parts of cancer patients (sometimes even unaffected individuals are targeted). It also includes multiomics information other than the genome" [8]. Information on *BRCA1/2* germline variants is useful for "treatment optimization, prognosis prediction, and onset prevention." Thus, interventions that include those who have not yet developed cancer can contribute to cancer prevention for the nation; therefore, it is desirable that MGPT be covered by health insurance.

References

1. Japanese Organization of Hereditary Breast and Ovarian Cancer. Guidelines for diagnosis and treatment of hereditary breast and ovarian cancer 2021. Tokyo: KANEHARA & Co., LTD; 2021.
2. Guidelines for the communication process in genomic medicine part 1: Focusing on comprehensive tumor genomic profiling [Revised 3rd edition]. 2021. http://sph.med.kyoto-u.ac.jp/gccrc/pdf/k102E_guidelines_part1_210908.pdf
3. Comprehensive tumor genomic profiling: materials for review of secondary findings, Ver 1.0. 2021. http://sph.med.kyoto-u.ac.jp/gccrc/pdf/k101E_kentousiryo_v1.pdf
4. Domchek SM, Friebel TM, Singer CF, Evans DG, Lynch HT, Isaacs C, et al. Association of risk-reducing surgery in BRCA1 or BRCA2 mutation carriers with cancer risk and mortality. JAMA. 2010;304:967–75.
5. Kurian AW, Ward KC, Hamilton AS, Deapen DM, Abrahamse P, Bondarenko I, et al. Uptake, results, and outcomes of germline multiple-gene sequencing after diagnosis of breast cancer. JAMA Oncol. 2018;4:1066–72.

6. Hirasawa A, Imoto I, Naruto T, Akahane T, Yamagami W, Nomura H, et al. Prevalence of pathogenic germline variants detected by multigene sequencing in unselected Japanese patients with ovarian cancer. Oncotarget. 2017;8:112258–67.
7. Kurian AW, Ward KC, Abrahamse P, Bondarenko I, Hamilton AS, Deapen D, et al. Time trends in receipt of germline genetic testing and results for women diagnosed with breast cancer or ovarian cancer, 2012-2019. J Clin Oncol. 2021;39:1631–40.
8. https://www.mhlw.go.jp/content/10901000/000341609.pdf

Current Status and Future Issues of the Activities of the Relevant Organizations

Makiko Dazai

Abstract When hereditary breast and ovarian cancer (HBOC) treatment in Japan began to gain momentum, there was a desire to have genetic testing, risk-reducing surgery, and PARP inhibitors covered by national health insurance, and, at the same time, there was a need for activities to promote correct understanding of HBOC among patients and the public. Furthermore, in order to standardize not only HBOC but also the treatment of hereditary tumors in general, it was necessary to strengthen the system for general education on cancer and genetics for all public, including those involved in the disease, and for efforts to solve social issues associated with genetic medicine. To this end, we believe that the role of patient and public involvement (PPI) was significant. In other words, the role of patient and public participation in identifying problems and understanding of research and investigation by providing opportunities to exchange opinions with patients and the general public was significant and helped to strengthen the HBOC treatment system. In addition, as expectations for HBOC treatment grow, we feel that it is essential to further strengthen support for the social background and family environment of each client, women's health care, and measures for genetic relatives (unaffected *BRCA* mutation carriers, at risk). In this chapter, we report on the seminars and the publication of a guidebook for patients and the general public organized by the Japanese Ministry of Health, Labor, and Welfare (MHLW), Research Group (preemptive medical care system for hereditary tumors using genomic data) from December 2020 to March 2022, with the aim of standardizing the treatment of HBOC and promoting the awareness and education of hereditary breast and ovarian cancer patients and the general public on hereditary tumors (especially HBOC). The publication of the guidebook for patients and public will be reported. In addition, we would like to report on the current status and issues of HBOC treatment obtained through support

M. Dazai (✉)
Specified Nonprofit Corporation Clavis Arcus, Association of Patients for Hereditary Breast and Ovarian Cancer Syndrome, Tokyo, Japan

Genetic Alliance JP, Shibuya-ku, Tokyo, Japan
e-mail: maki@geneticalliance.jp

D. Aoki et al. (eds.), *Practical Guide to Hereditary Breast and Ovarian Cancer*,
https://doi.org/10.1007/978-981-99-5231-1_9

activities for patients, as well as the social issues associated with these issues that need to be resolved in the future.

Keywords Hereditary breast and ovarian cancer (HBOC) · *BRCA1/2* pathologic variant carriers · Unaffected carriers · HBOC treatment system · Patient and public involvement (PPI) · Shared decision-making (SMD)

1 Introduction

It was not the message sent to the world by an international actress in 2013, but the partial coverage by National Health Insurance of Hereditary Breast and Ovarian Cancer (HBOC) treatment, which came into effect in April 2020, that made a big movement in HBOC treatment in Japan and made it widely known to the people concerned and public. The fact that the measures for people who have already developed cancer are being established as routine medical treatment and the strengthening of the medical care delivery system was also strongly felt. Furthermore, the change to the fact that PARP inhibitors and companion diagnostics have become offered and affordable medical care is so great that it is far different from the level of awareness when they were first covered by national health insurance. It is undeniable that a large percentage of patients who were reluctant to undergo genetic testing or risk-reducing surgery in the past, even though it was recommended to them, now have to wait for their turn to undergo risk-reducing surgery, which has a significant impact on reducing their financial burden. In addition, the presence of *BRCA1/2* genetic testing and risk-reducing surgery among the treatment options proposed by attending physicians has provided patients with a great sense of security.

In the past, even though HBOC measures in Japan were based on research and treatment in private practice, the generation that accepted HBOC in consideration of reducing the incidence of cancer and the impact on genetic relatives, and the generation that was diagnosed under insurance treatment, may have different ways of accepting HBOC. For this reason, in an age when more and more patients can choose, we believe that everyone needs to have an equal right to know about HBOC. To this end, it is necessary and required to continuously provide information that cannot be conveyed in daily medical care, and to increase opportunities for sharing and updating information, so that information can reach those who may be at risk.

It is expected that someday it will be commonplace for everyone to know their physical, which will lead to health management in control of cancer and other diseases and will play a major role in decision-making support in line with one's lifestyle.

2 Awareness and Educational Activities for the Public and HBOC Patients

It is essential that information be provided by experts in the preemptive medical care delivery system. Under the Health and Labor Sciences Research Grants Program, a system of regular public lectures has been established to educate patients and the public. A series of public lectures titled "What we currently want to talk and think about" [Ima, tsutaetai-koto, kangaetai-koto] was held online and open to anyone, including patients, the public, and students, free of charge. Since their inception, the lectures have been attended not only by the general public but also by genetic counselors, nurses, and other medical professionals and have been used not only to raise interest in HBOC treatment but also as a venue for learning basic knowledge necessary for actual medical practice.

What We Currently Want to Talk and Think About [Ima, tsutaetai-koto, kangaetai-koto] 2020–2022 [1]

1. Cancer and Genes/Genetics 2020.12

   Akihiro Sakurai (Sapporo Medical University School of Medicine)

   The basics of cancer, genetics, and heredity were described in an easy-to-understand presentation. He explained the differences between people according to their genes, the importance of knowing that cancer is hereditary, and what measures can be taken with current medical care by sharing genetic risks not only with oneself but also with one's family.
2. Hereditary Breast and Ovarian Cancer and Risk-Reducing Salpingo-oophorectomy (RRSO) 2021.2

   Yusuke Kobayashi (Keio University School of Medicine).

   HBOC from a gynecologist's perspective.

   A detailed explanation of risk-reducing salpingo-oophorectomy (RRSO). Possible changes to the body that can occur with removal of the ovaries, fallopian tubes, and uterus. Answers questions about the length of hospital stay and cost required for risk-reducing surgery. We have structured this lecture not only for patients alone but also for patients with their family members or their partners as well as other lectures.
3. How to Use Genetic Counseling in Hereditary Breast and Ovarian Cancer (HBOC). 2021.3

   Noriko Tanabe (National Cancer Center Hospital).

   Patients and their family of genetic counseling before undergoing genetic testing and in various aspects of treatment and cancer genome medicine. When they hear the word "counseling," they feel resistance, do not know what to ask for, and have other concerns. To learn about those concerns and questions, we asked him to explain about genetic counseling and certified genetic counselors.
4. Knowledge Is Power, Making the Right Choice for You. Risk-Reducing Mastectomy (RRM) 2021.4

   Hideko Yamauchi (St. Luke's International Hospital).

The fourth issue of this series explained risk-reducing mastectomy (RRM) from the perspective of a breast doctor.

The fact that people who are genetically more likely to develop breast cancer have the option of having their breasts removed before the onset of the disease. She shared the latest information to overcome anxiety, consider what is known and what is not known to prevent cancer before it develops, and to make the right choice with the right knowledge.

Patients need accurate information from professionals such as doctors. Among the patients who received information from specialists at this lecture, many of them started consulting with their doctors and actually implemented RRM.

5. Measures Unaffected *BRCA* Mutation Carrier 2021.6.

   Issei Imoto (Aichi Cancer Center)

   Starting in April 2020, preventive treatment for HBOC will be partially covered by national health insurance, and the treatment system at facilities and the treatment costs borne by patients are changing significantly.

   With the increase in the number of cancer patients with a confirmed diagnosis of HBOC, more and more of their genetic relatives, mainly their siblings and children, are receiving genetic counseling to learn about their risk and considering and conducting genetic testing while they are "unaffected" by the disease. Does knowing if you are an unaffected *BRCA* mutation carrier (cancer-prone person) before you develop cancer help you manage your health? Is not it burdensome just in terms of feelings, cost, time, etc.? These questions were discussed with the audience.

   We could also learn about the difference between *BRCA1/2*-related genetic testing and cancer genomic medicine at the lecture.

6. About Me, and What You Tell My Family 2021.8.

   Aya Kanno (Clavis Arcus; Association of Patients for Hereditary breast and ovarian cancer syndrome)

   In the past, the lectures were mainly conducted by medical professionals, but this time, a patient shared her thoughts and experiences about cancer and hereditary cancer from her personal point of view. She talked about her experience of sharing her cancer with her family and her newfound awareness about HBOC.

   As an HBOC patient, she shared what she felt when she was informed twice about cancer and when she came to know she had HBOC. She talked about her experiences of telling her daughter and sister about the risk-reduced operation, her awareness of HBOC after a lot of experiences, and her thoughts on HBOC these days. Telling stories of one's experiences is precious, and they give courage to the listeners. Some patients even decided to start talking about HBOC with their families because of the stories they heard.

   As for informing one's children of HBOC, many people worry about how to tell and when to tell in terms of their age and timing. Moreover, this is the most talked about topic in peer support. On the other hand, children often want their parents to be honest about their cancer and genetics.

7. (1) Cancer and Genes/Genetics

Akihiro Sakurai
(2) Broadening our Understanding Will Support us.
Masahiro Kawakami
It is important to know about heredity and hereditary cancer. By knowing, we can aim at a society where we think about each other and care and respect each other. We could know how genome medicine and genome analysis have progressed based on research shared in public lectures.

Other open seminars for medical professionals include "The Role of MRI in HBOC Practice and Breast Cancer Surveillance" in March 2021 and "Seminar to Explain the 2021 Edition of the Guidelines for Hereditary Breast and Ovarian Cancer Treatment" in August 2021.

The videos are still available free of charge on YouTube and the research group's website. (https://www.geneticsinfo.jp).

3 Cosponsoring and Cooperating With Related Seminars to Strengthen Public Awareness Efforts

The "Insight" seminar was cohosted by organizations involved in genome medicine to provide a forum for sharing information and discussing solutions to social issues in parallel with the accelerating pace of genomic medicine and HBOC treatment.

Through talk sessions on the theme of legislation to eliminate discrimination and prejudice caused by genetics and easy-to-understand presentations to the general public on how to view and understand the results of research targeting, unaffected *BRCA* mutation carriers have helped to promote understanding of genome information management and research.

4 Guidebook for Treatment and Understanding of HBOC: PPI and Products

In July 2022, the MHLW Science Research Group (Akihiro Sakurai) and Japanese Organization of Hereditary Breast and Ovarian Cancer (JOHBOC) (Seigo Nakamura) published Japan's first guidebook for patients and public of hereditary breast and ovarian cancer, "Hereditary Breast and Ovarian Cancer (HBOC) Guidebook for the public 2022".

The content is in accordance with the 2021 edition of the Hereditary Breast and Ovarian Cancer (HBOC) Clinical Practice Guidelines, which were published in 2021, and explains the actual HBOC treatment as of 2022 in a 59-Q&A format, which is divided into 3 chapters plus a collection of columns and documents. The Q&A includes the actual voices of doctors, nurses, certified genetic counselors, and HBOC patients. The book is aimed at HBOC patients, family members, genetic relatives, the general public, medical professionals, and genetic counselors connected to HBOC-related cancers (breast cancer, ovarian cancer, prostate cancer, pancreatic cancer, malignant melanoma, etc.) and is written in a simple, easy-to-read format. It also includes a list of websites with essential information, a glossary of terms, and a list of coverage of National Health Insurance. It is designed for three generations of parents, siblings, and children to learn together.

4.1 Content and Typical Questions [2]

Chapter 1: What you need to know about HBOC.

Q1: What is HBOC?

Q2: If I am diagnosed with HBOC, what type of cancer am I likely to develop?

Chapter 2: What do I need to know if I am diagnosed with HBOC?

Q20: What are my options and choices?

Q21: What is surveillance? Is it different from a Cancer screening?

[Breast cancer]

Q29: What are the characteristics of breast cancer caused by HBOC?

Q30: What about breast self-examination (self-palpation)?

[Ovarian cancer]

Q40: Are there any characteristics of ovarian cancer (fallopian tube cancer and peritoneal cancer) that develop with HBOC?

Q41: Please tell me about risk-reducing salpingo-oophorectomy (RRSO)

[Pancreatic cancer, prostate cancer, and other cancers].

Q48: Are there any characteristics of pancreatic cancer that develop in HBOC?

Q49: Is surveillance for pancreatic cancer necessary? How does surveillance take place?

Chapter 3 What you need to know to watch out for in daily life.

Q53: Are there any lifestyle habits I should keep an eye on?

Q54: Is there anything I need to avoid or be aware of in my daily life?

Columns, List of HBOC Information

*Information Website.

*List of HBOC Public health Insurance Coverage.

*Glossary and Abbreviations.

4.2 For Creation and Editing

In creating this book, we intended to share what patients and their families would like to know, what their concerns are, and what explanations they would like to have from medical professionals. By sharing these, we considered that this book would be of benefit to those receiving HBOC treatment. We conducted a preliminary questionnaire survey for people who are about to receive HBOC medical care and for people who have already been through the HBOC treatment on what they would like to know and what they wanted to know when they have had the HBOC treatment (request for a questionnaire on the development of HBOC clinical practice guidelines). At the same time, the team of developers, who are medical professionals, collected questions frequently asked by patients and their families as well as information that they always explain during medical treatment.

Therefore, after narrowing down the questionnaires, we have launched full-scale Patient and Public Involvement (PPI) based on the Q&A questions.

The development team consists of the overall leader, Yusuke Kobayashi, Department of Obstetrics and Gynecology, Keio University School of Medicine, and four area leaders who are experts in their fields. The development committee is composed of 7 people: a nurse, a certified genetic counselor, a scientific expert, and HBOC patients. In addition, there are 14 coauthors and 3 advisors.

4.3 Introduction to Patient and Public Involvement (PPI)

What was most important in creating this guidebook was that patients and the public would read it and use it as a means for getting information about cancer, gene mechanisms, and tools for decision-making support related to diagnosis of HBOC. The guidebook is not only for patients but also anyone to learn about hereditary breast and ovarian cancer.

As I mentioned earlier, clinical practice guidelines for medical professionals have already been published, and although the people concerned have been following them, it was assumed that it would be difficult for the general public to understand them.

For cancer treatment, it is important that patients and their families correctly understand hereditary breast and ovarian cancer, and we value shared decision-making, in which medical professionals and patients work together to provide medical care. In order to increase understanding, we decided to introduce Patient and Public Involvement (PPI) from the planning stage, which is what we have done.

In the implementation process, three members of the PPI project team (Yusuke Kobayashi, Kenta Masuda, and Makiko Dazai) and a total of 82 HBOC patients, *BRCA1/2* unaffected carriers, non-HBOC breast cancer patients, patients with other types of cancer, family members of patients with genetic diseases, and the general public participated on the PPI team and worked together. Selecting the participants for this book, we considered not only the mere classification of HBOC and diseases, but also the positions and diverse ways of thinking in daily life, including those such as homemakers, educators, doctors, nurses, pharmaceutical companies, general company workers, freelancers, and other occupations and attributes.

Before the exchange of opinions, some of the participants had no experience in patient/public participation, so we invited a lecturer to explain PPI and establish the roles and objectives of participation.

The commitment for the meeting for exchanging opinions was 4 times (10 hours). In fact, considering that all of the approximately 250 questions provided as preliminary materials and the completed manuscript (answers) were reviewed, it cannot be denied that it placed a burden on those who participated. However, from the process in which the author and the development team reviewed and improved each question, proposal for revision, and opinions on wording, we were united in our desire that the book will be picked up by a large number of people, and they will have a correct understanding of HBOC and move forward together.

It is not an exaggeration to say that there has never been a medical guidebook that has been completed through the participation of so many public. It is a memorable book produced by the joint effort of medical professionals and the public. I hope that many people will deepen their understanding of hereditary breast and ovarian cancer through this book and use it as a guide for the treatment of cancer.

5 Other Research Projects

5.1 Genetic Cancer Support Program(GCSP)/HBOC Peer Support [3, 4]

Proper diagnosis of HBOC is very important because it may affect the treatment and postsurgery surveillance methods. In addition, since the genetic constitution may be shared by genetic relatives such as brothers, sisters, and children, it is essential to improve the quality of HBOC medical care including genetic counseling. By properly diagnosing not only those who have developed cancer, but also those who have the predisposition to HBOC but have not developed cancer, and by providing testing for early detection and treatment to reduce the risk of developing cancer in addition to optimal treatment, the best personalized medicine can be provided for both patients and their families. In order to provide such support to patients and their families, peer support is essential for sharing feelings and experiences that are inaccessible to medical professionals.

Peer support through various patient support groups, patient discussion groups in hospitals, cancer consultation support centers, and medical care providers have been implemented in many places in Japan, and the promotion of peer support is called for in the Plan for the Promotion of Cancer Control.

Through this course Genetic Cancer Support Program (GCSP), those diagnosed with HBOC can work together to think about how they can share their experiences and information, and how to maximize the effectiveness of peer support and its role in providing high quality mutual support, based on their experiences. The GCSP aims to build a mutual support system that builds a secure and trusting relationship with one another by educating themselves on accurate basic knowledge of heredity and cancer, mutual support for patients, and peer counseling, so that they can provide peer support.

5.2 *Future Genetic Medicine That HBOC Patients and Their Families Want*

Through these activities, it has become possible to provide up-to-date and accurate information to patients, their families, and medical professionals. However, the number of people diagnosed with HBOC has been increasing rapidly. In the current situation, medical professionals need to provide more correct information and medical support for those affected, and ongoing awareness and support activities will be essential.

In the future, it will also be essential to raise awareness among men and take measures and support for women's health care after cancer and risk-reducing surgery. It is essential to develop laws against discrimination caused by heredity, since discrimination has a psychological impact,

And in terms of school education, advice from medical professionals and support for solutions are crucial.

In order to ensure that people can receive HBOC medical treatment with peace of mind wherever they are, we will create opportunities for interaction and exchange of ideas with people in various positions, establish a support system.

I think that we should implement initiatives that are considerate of "people in society, environment and medical care", sustainable and safety-oriented in all across Japan.

Finally, looking back on the history of HBOC treatment, it gives us hope that we have walked the path together with medical professionals and researchers from the time when it was all private practice to the time when it is partially covered by health insurance. It has been a big support.

Guidebooks that could not be found 10 years ago are now available to everyone, and information on HBOC sought by patients and the public is regularly released, creating an environment in which questions can be answered. Our next agenda is further advance of our understanding of social issues and the problems of unaffected carriers who are not yet covered by health insurance, which is for ourselves and for our next generation.

Not only do I feel sad when I learn that I have hereditary cancer, but I also feel my future is bright because of knowing that I have hereditary cancer. No matter how painful the treatments have been, I am proud of myself as an HBOC survivor. The reason why I feel like that is that I am grateful that there are medical professionals in front of me who are willing to go ahead, reach out, and lift me up, even when the barriers are insurmountable.

6 About "Clavis Arcus"

Clavis Arcus, the first and only patient association supporting *BRCA1* and *BRCA2* pathogenic variant carriers and their families in Japan. The organization was established in 2014 and certified as a nonprofit organization by the Tokyo Metropolitan

Government in 2015. The organization aims to provide a gathering space for the members to support each other and to deepen knowledge and understanding of hereditary tumors. There are 145 members among Japan and has a branch in Pennsylvania, US.

They organization provides consultations by phone, e-mail, and in person as well as holding patients' gatherings. We started the "Institute of Genetic Studies" for further understanding of hereditary cancers, education for peer supporter, and holds Learning about Genetics for Families seminars annually.

Recently, photo panel exhibitions is running nationwide in Japan. The photos consist with the image of the members themselves and letters from their family [3] (http://www.clavisarcus.com)

7 About "Genetic Alliance JP"

Genetic Alliance JP was established through the alliance and cooperation of patients suffering from hereditary disorders, their families, and related organizations to materialize and spread appropriate genomic medicine through mutual support, education and awareness, policy proposals, research and study, health care, and welfare improvements activities as well as to resolve hereditary cancer patients' social issues. To date, 14 organizations related to genomic medicine are part of this alliance [5] (https://www.geneticalliance.jp).

References

1. The Japanese Ministry of Health, Labour and Welfare (MHLW), Research Group (Preemptive medical care system for hereditary tumors using genomic data). In: Hereditary Breast and Ovarian Cancer information 2020–2022 (https://geneticsinfo.jp) and Public lectures, what we currently want to talk and think about [Ima, Tsutaetai-koto, Kangaetai-koto] 2020–2022.
2. The Japanese Ministry of Health, Labour and Welfare (MHLW), Research Group (Preemptive medical care system for hereditary tumors using genomic data) Research Group in Health and Labour, preemptive medical care system for hereditary tumors using genomic data and Registration Committee of the Japanese Organization of Hereditary and Breast and Ovarian Cancer (JOHBOC). PPI: patient and public involvement project 2021–2022 and guidebook for treatment and understanding of HBOC 2022. Tokyo: Kanehara Co. Ltd.; 2022.
3. Specified Nonprofit Corporation Clavis Arcus; Association of Patients for Hereditary breast and ovarian cancer syndrome, Tokyo, Japan. https://www.clavisarcus.com.
4. The Japanese Ministry of Health, Labour and Welfare (MHLW), Research Group (Preemptive medical care system for hereditary tumors using genomic data), Clavis Arcus, In: Genetic Cancer Support Program (GCSP) for HBOC Information book 2021–2022.
5. Dazai M, et al. Genetic Alliance JP information, In: GENETIC LEAGUE 2019.11. Tokyo: General Incorporated Association Genetic Alliance JP; 2017.

Part V
Congress Award

Significance of Understanding HBOC for At-Risk Relatives in Prostate and Pancreatic Cancer Patients Who Tested Positive on the Germline *BRCA1/2* Genetic Testing in the Current Japanese Healthcare System

Eriko Takamine

Abstract Germline *BRCA1/2* genetic testing for prostate and pancreatic cancer patients who meet certain criteria was recently approved in the Japanese healthcare system as a companion diagnostic test to see the eligibility for poly (ADP-ribose) polymerase (PARP) inhibitors, olaparib. When a pathogenic variant is identified through the test, it indicates that they are eligible for olaparib, but this also confirms that they have hereditary breast and ovarian cancer (HBOC). Once HBOC is confirmed, it becomes important to consider utilizing that information for their family. This chapter discusses perspectives including the importance of accurate understanding of HBOC, influences on their at-risk relatives, and communication with the relatives of prostate and pancreatic cancer patients.

Keyword HBOC · Prostate cancer · Pancreatic cancer · At-risk relatives · Genetic test · *BRCA1/2* · BRACAnalysis

E. Takamine (✉)
Department of Medical Genetics, Tokyo Medical and Dental University Hospital, Tokyo, Japan

Department of Precision Cancer Medicine, Tokyo Medical and Dental University Hospital, Tokyo, Japan
e-mail: takamine.canc@tmd.ac.jp

D. Aoki et al. (eds.), *Practical Guide to Hereditary Breast and Ovarian Cancer*, https://doi.org/10.1007/978-981-99-5231-1_10

1 Introduction

Hereditary breast and ovarian cancer (HBOC) is one of the most common hereditary cancer predisposition syndromes. Breast cancer and ovarian cancer are major symptoms of HBOC, but other cancers such as prostate cancer and pancreatic cancer are known to be associated with it as well [1, 2]. Even though prostate and pancreatic cancers are not emphasized as much as it is with breast and ovarian cancer, it is essential to focus on them as well. Statistically, prostate cancer is the most common cancer among males in Japan [3]. The number of newly diagnosed prostate cancer patients was 92,021 in 2018 and that of those who died with it was 12,759 in 2020 [3]. Of those numbers, pancreatic cancer represented 42,361 and 37,677, respectively [4]. In Japan, HBOC is reported to be found in about 1.2% of prostate cancer patients [5] and about 3.4% of pancreatic cancer patients [6].

HBOC is caused by a pathogenic or likely pathogenic variant (both together referred as pathogenic variants below) in the *BRCA1* and *BRCA2* genes [1, 2]. Since causative genes are identified, genetic testing is used to diagnose it.

With clinical experiences and perspectives as a certified genetic counselor, the author discusses the significance of understanding HBOC for at-risk relatives of prostate and pancreatic cancer patients who are diagnosed as HBOC in the current Japanese healthcare system.

2 Brief History of Germline *BRCA1/2* Genetic Testing in the Japanese Healthcare System

2.1 Prior to the Public Health Insurance Coverage

Germline *BRCA1/2* genetic testing was performed only without public health insurance until 2018. In other words, patients had to pay the cost on their own which was roughly JPY 200,000 to 300,000 for testing the two genes. It was normally recommended to those who had breast or ovarian cancer and met certain criteria such as the NCCN guidelines® for HBOC. Genetic counseling was provided to help them understand HBOC and discuss pros and cons of taking the genetic testing. However, the cost hindered patients from taking it. Additionally, patients are required to pay the fee of pre–genetic counseling and post–genetic counseling sessions which are not covered by the insurance. Even if patients were diagnosed as HBOC, almost all following actions such as active surveillance of cancer, risk-reducing salpingo-oophorectomy (RRSO) and risk-reducing mastectomy (RRM) had to be done without the insurance. Also, there was no medicine approved specifically for HBOC. At that time, patients with prostate or pancreatic cancer were not actively targeted by the genetic testing.

2.2 Insurance Coverage for Breast Cancer and Ovarian Cancer

Germline *BRCA1/2* genetic testing was first covered by the public health insurance with the revision of medical fee (the ministry of health, labor, and welfare) in 2018 as a companion diagnostic test (CDx) of olaparib, poly (ADP-ribose) polymerase inhibitors, for HER2 negative metastatic or recurrent breast cancer. There were multiple genetic testing companies, but BRACAanalysis® provided by Myriad Genetics, Inc. was the only test that was approved with the insurance. In the following year, the testing was covered as CDx of olaparib for maintenance treatment of platinum-sensitive ovarian cancer after first chemotherapy.

The cost of the testing was set at JPY 202,000; however, patients' financial burden has decreased to 30% of it at most due to the insurance coverage. Along with that, the post-test genetic counseling session started to be covered at registered institutions of genetic counseling only if pre-test counseling had been provided. Some institutions which did not meet the genetic counseling requirements still had to provide it for free or at patient's own expense.

2.3 Insurance Coverage for HBOC

In 2020, the germline *BRCA1/2* genetic testing for breast cancer patients who meet the criteria in Fig. 1 and the ovarian cancer patients is covered by the insurance for the diagnosis of HBOC [7], which was the first approval not as CDx. Along with the HBOC diagnosis, prophylactic surgeries such as RRSO and RRM, breast reconstruction, and breast surveillance with MRI are covered. This was the first approval to resect organs that have not developed any symptoms. However, this approval increased the number of at-risk relatives who are completely healthy and not eligible for the genetic testing with the insurance.

Fig. 1 Requirements to take the germline *BRCA1/2* genetic testing for the diagnosis of HBOC in the current Japanese healthcare system [7]. Breast cancer patients need to meet at least one of the requirements. Any type of ovarian cancer patients is eligible for the testing

Germline *BRCA1/2* genetic testing requirements for HBOC:

(At least one of the following needs to be met)

- Breast cancer
  - Age of 45 or younger
  - Triple negative breast cancer at age of 60 or younger
  - More than one primary breast cancer
  - At least one breast or ovarian cancer patient within third-degree relatives
  - Male breast cancer
- Ovarian cancer

2.4 *Insurance Coverage for Prostate Cancer and Pancreatic Cancer*

At the end of 2020, the germline *BRCA1/2* genetic testing was approved to be covered by the insurance for metastatic castration-resistant prostate cancer (mCRPC) and unresectable pancreatic cancer as CDx of olaparib. For prostate cancer, somatic *BRCA1/2* genetic testing also became available since among those who have germline or somatic *BRCA* variants, roughly half of them is known to be somatic [8, 9] and both of them are eligible for olaparib. For pancreatic cancer, however, germline testing was the only approved test. In the history of HBOC practice in Japan, these two cancers were not discussed as often as breast and ovarian cancers. Hence, some considerations need to be made when it comes to practice which will be stated in the next section.

3 Prostate Cancer Patients and HBOC

3.1 *Background of Those Who Take the Germline* **BRCA1/2** *Testing*

According to the case-control study conducted by Momozawa et al., the mean age at diagnosis of prostate cancer without germline pathogenic variants was 71.0 and that of with pathogenic variants was 69.0 which is 2.0 years younger, although this study targeted six other genes such as *HOXB13* and *ATM* as well [5]. That is, when they are diagnosed as HBOC through CDx, which cannot be conducted until they have mCRPC, many of them are assumed to be in their 70s and 80s. Although patients are at least briefly explained about HBOC before taking the germline *BRCA1/2* genetic testing, their main interest is eligibility of olaparib. In addition, it is indicated that prostate cancer with a germline *BRCA2* variant has a higher Gleason score and is aggressive [5].

Typically, in most cases, patients' primary physicians order germline *BRCA1/2* genetic testing, and when it comes back positive, they send patients to genetic counseling. On the other hand, some institutions send all patients to genetic counseling before the test. The flow of the test varies depending on the institution.

3.2 *Patients' Accurate Understanding of HBOC*

The name HBOC could potentially cause misunderstanding of the genetic condition since the name has breast and ovaries but not prostate, giving the patients the impression that it is not applied to prostate cancer patients or males. Therefore, it is essential to emphasize that it affects both males and females and recommend them to go to genetic counseling at least once to understand the whole picture. Patients who normally see their physicians alone and hear the positive results might not be aware of the significance of HBOC. Furthermore, not all HBOC family has family history of cancer, especially when the family size is limited or a variant is inherited from the paternal side. In addition, aging and hearing problem could come into play when understanding. Therefore, genetic counseling providers may need to pay extra attention to that aspect of them. Some patients may want to avoid talking about genetic inheritance due to potential cultural and social stigma or prejudice.

3.3 Influences on Family

When diagnosed as prostate cancer in their 70s and 80s, most of their children, nieces and nephews are already grown up if they have them. In addition, it is not rare to have grandchildren. Therefore, emphasizing on familial influence plays an important role when explaining about HBOC to the patients.

For example, an 80-year-old man just found out he has a *BRCA2* pathogenic variant. As drawn in Fig. 2, he (II-2) has two daughters who are 49 (III-4) and 46 (III-6) years old and they also have children (IV-7, 8, 9, 10, 11). Since HBOC is inherited in the manner of autosomal dominant, each of his daughters has a 50% of chance of having the same variant. They have not developed any cancer at this point, but it does not eliminate the chance.

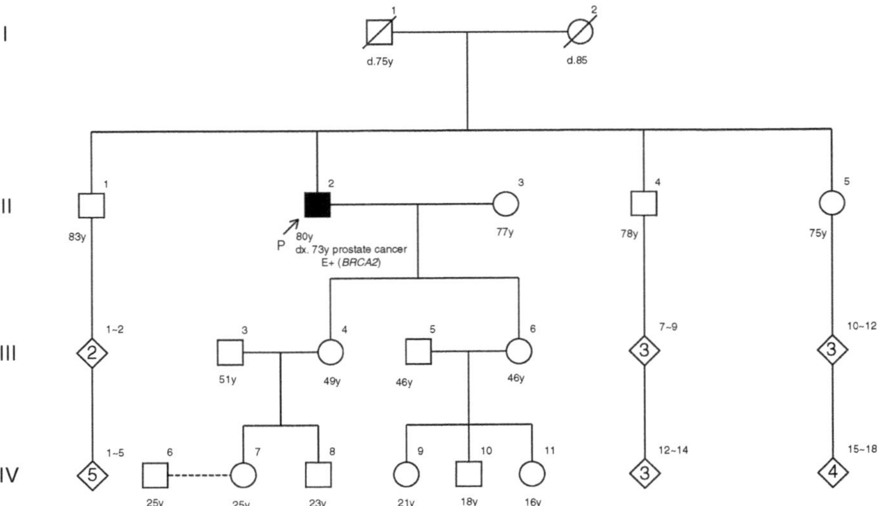

Fig. 2 An example family tree of a prostate cancer patient with a germline *BRCA2* pathogenic variant. The patient (II-2) who is colored is the proband (P) and diagnosed as prostate cancer. He has two daughters (III-4, 6). He also has an older brother (II-1), a younger brother (II-4), and a younger sister (II-5). None of them has developed cancer at this point

3.3.1 Genetic Counseling with Family

The main focus for the prostate cancer patients is to treat cancer, so their main interest would be using olaparib. Even if they understand HBOC, they might not perceive there is an impact on themselves due to their age or the fact they have already developed cancer. Therefore, the significance of understanding genetics and HBOC could be disregarded. Also, having advanced cancer may not give them enough time to think. Considering the situation, physicians and other medical professionals should advise them to go to genetic counseling with their family relatively soon. In the case of Fig. 2, his wife (II-3) and/or his daughters (III-4, 6) could be potential companions. His daughters are in their 40s, which means they have already reached the age of developing HBOC-related cancer and should undergo appropriate surveillance if they also have the pathogenic variant.

Some patients could be hesitant telling the fact of having HBOC to their children for various reasons. One possible reason is that they are worried that they have inherited the variant from them which causes guilt toward their children, and have not realized the significance of potential actions for at-risk relatives such as surveillance, RRSO, and RRM. If patients are still not willing to share the information with their family after learning about HBOC, its inheritance pattern, and actionability, their opinions need to be respected. However, that is not the end of genetic counseling. They may change their opinions in the future, so approaching them occasionally is recommended.

In Fig. 2, if the patient's daughters learn about themselves being at risk, they might want to talk with their partners and, if possible, their children. However, one of the children (VI-7) may face a conflicting situation since she is engaged with her partner, and it might be hard for her to decide when she would like to tell him about HBOC that runs in her family.

3.3.2 Communication With Extended Family

In Fig. 2, the patient (II-2) has two brothers and a sister who are in their 70s and 80s and have not developed cancer. Ideally, he tells them about HBOC and they would visit genetic counseling nearby to get more information about it. The chance of them not having the variant might be higher because they have not developed cancer up until this age. Nonetheless, it still does not eliminate the chance since the penetration rate of HBOC is not 100% [2]. Also, they have children and grandchildren, so surveillance and prophylactic surgeries need to be considered as appropriate if they also have the variant. However, communication with extended family could be challenging depending on the relationships so cautions and advice from genetic counseling providers may be needed.

4 Pancreatic Cancer Patients and HBOC

4.1 Background of Those Who Take the BRCA1/2 *Testing*

As stated, the number of pancreatic patients has been increasing, and in 2018, the number of newly diagnosed patients was 42,361, which included both males and females [4]. The number of deaths was 36,677 in 2020, and the 5-year relative survival rate of those who were diagnosed between 2009 and 2011 was 8.5% [4]. As the statistics shows, the prognosis of pancreatic cancer is poor. The mean age at diagnosis is 67 for all pancreatic cancer patients, and unlike other HBOC-related cancers, carrying a pathogenic variant does not seem to affect the age at diagnosis [6]. If pancreatic cancer patients with a pathogenic variant have children, most of their children are likely in their 20s to 30s, possibly in 40s when the patients are diagnosed with HBOC.

4.2 Accurate Understanding of HBOC

As is the case with prostate cancer, the patients need to understand what HBOC is accurately, although there is no "pancreas" in the name of the disease. When patients' physical conditions worsen, treating cancer with olaparib may become the only focus they can handle at the time, leaving the understanding of HBOC and genetics behind. Considering the situation, support from their family especially their partners play a part in communicating about HBOC to at-risk relatives. Therefore, it is recommended to bring at least one relative when they hear results of germline *BRCA1/2* genetic testing.

4.3 Influences on Family

Like prostate cancer patients, it is important for pancreatic cancer patients to visit genetic counseling with their family and, as needed, communicate with their extended family. However, children of pancreatic cancer patients who just found out to have a *BRCA* pathogenic variant through CDx are more likely to be younger than those of prostate cancer due to the patients' age of diagnosis. Since most of them are assumed to be in their 20s and 30s, some may be facing major life events such as marriage and pregnancy. Hence, conflicts whether to tell them about HBOC now or later could possibly arise.

If female children have the pathogenic variant, they may be already at the age to start the surveillance for breast and ovarian cancers. Regardless of their biological sex, they are recommended to start surveillance for the pancreas after the age of 50 years since they have a first-degree family history of pancreatic cancer [10].

Pancreatic cancer screening is not actively performed for healthy individuals [10]. Therefore, it is ideal to go to genetic counseling occasionally or have a physician who oversees them so that when they reach the age, they can start the surveillance. However, healthy individuals are not covered by the insurance to receive the surveillance as of now, so cost could be challenging depending on the individuals.

5 Roles of Genetic Counseling Department

What has been described above about prostate and pancreatic cancer patients is crucial and needs to be considered before and when genetic counseling is provided. Nonetheless, patients are not the only ones who battle the genetic conditions. Medical professionals also battle building an environment where they can communicate with other related departments to provide appropriate medical service and information at the right time to the patients. To be specific, physicians and genetic counseling providers such as medical geneticists and genetic counselors are recommended to have a meeting on a regular basis to get to know the role of each department and exchange most recent information to establish HBOC practic as a whole.

Once patients visit genetic counseling, it is an important step forward to potential utilization of the genetic information in the family. A family tree would be drawn, and at-risk relatives would be identified. Patients are not forced to talk to their family if they do not wish to do so. The pros and cons of disclosing HBOC to relatives such as surveillance and the possible psychosocial burden are discussed. If they could guide relatives to genetic counseling, the relatives may take advantage of the genetic information. The relatives' desire and willingness to take the genetic testing should be respected. Even if they do not desire to take it soon, information about genetic counseling should be provided, so that they at least know where to go for testing or more information in the future.

The genetics department is in the position where they take the lead within an institution to build a team for HBOC practice as they keep track of patients' and their relatives' health management situations and familial communications comprehensively.

6 Conclusion

Since the introduction of germline *BRCA1/2* genetic testing as CDx for prostate and pancreatic cancer patients in 2021, some of them are diagnosed as HBOC along with the eligibility of olaparib. At-risk relatives could benefit from the patients' accurate understanding of HBOC and utilize the information for their health management once they learn about HBOC through genetic counseling. In the long run, it can be beneficial to share the information with patients' relatives. Although the current insurance coverage does not cover healthy relatives' testing, surveillance, and prophylactic surgeries, knowing the genetic information could play an important role in the family especially for female offspring since many of them are assumed to have reached the recommended ages of surveillance. For the better HBOC practice, collaboration and communication among medical professionals such as primary physicians, medical geneticists, and genetic counselors will be essential.

References

1. Momozawa Y, Sasai R, Usui Y, Shiraishi K, Iwasaki Y, Taniyama Y, et al. Expansion of cancer risk profile for BRCA1 and BRCA2 pathogenic variants. JAMA Oncol. 2022;8(6):871–8. https://doi.org/10.1001/jamaoncol.2022.0476.
2. Antoniou A, Pharoah PDP, Narod S, Risch HA, Eyfjord JE, Hopper JL, et al. Average risks of breast and ovarian cancer associated with BRCA1 or BRCA2 mutations detected in case series unselected for family history: a combined analysis of 22 studies. Am J Hum Genet. 2003;72(5):1117–30. https://doi.org/10.1086/375033.
3. Prostate. National Cancer Center Ganjoho Service. https://ganjoho.jp/reg_stat/statistics/stat/cancer/20_prostate.html. Accessed 15 Jun 2022.
4. Pancreas. National Cancer Center Ganjoho Service. https://ganjoho.jp/reg_stat/statistics/stat/cancer/10_pancreas.html. Accessed 15 Jun 2022.
5. Momozawa Y, Iwasaki Y, Hirata M, Liu X, Kamatani Y, Takahashi A, et al. Germline pathogenic variants in 7636 Japanese patients with prostate cancer and 12 366 controls. J Natl Cancer Inst. 2020;112(4):369–76. https://doi.org/10.1093/jnci/djz124.
6. Mizukami K, Iwasaki Y, Kawakami E, Hirata M, Kamatani Y, Matsuda K, Endo M, Sugano K, Yoshida T, Murakami Y, Nakagawa H, Spurdle AB, Momozawa Y. Genetic characterization of pancreatic cancer patients and prediction of carrier status of germline pathogenic variants in cancer-predisposing genes. EBioMedicine. 2020;60:103033. https://doi.org/10.1016/j.ebiom.2020.103033.
7. CQ1. Guidebook for Diagnosis and Treatment of Hereditary Breast and Ovarian Cancer Syndrome 2017. Japanese Organization of Hereditary Breast and Ovarian Cancer (JOHBOC). http://johboc.jp/guidebook2017/toc/2-1index/cq1/. Accessed 26 June 2022.
8. Abida W, Armenia J, Gopalan A, Brennan R, Walsh M, Barron D, et al. Prospective genomic profiling of prostate cancer across disease states reveals germline and somatic alterations that may affect clinical decision making. JCO Precis. Oncologia. 2017;2017:PO.17.00029. https://doi.org/10.1200/PO.17.00029.
9. Lang SH, Swift SL, White H, Misso K, Kleijnen J, Quek RGW. A systematic review of the prevalence of DNA damage response gene mutations in prostate cancer. Int J Oncol 2019 Sep;55(3):597–616. https://doi.org/10.3892/ijo.2019.4842.
10. Pancreatic cancer FQ2. Guidelines for Diagnosis and Treatment of Hereditary Breast and Ovarian Cancer 2021. Japanese Organization of Hereditary Breast and Ovarian Cancer (JOHBOC). http://johboc.jp/guidebook_2021/doc2-5/e3/. Accessed 19 Jun 2022.

Handling Germline Findings in Ovarian Cancer Cases

Mika Okazawa-Sakai

Abstract Comprehensive cancer genomic profiling (CGP) can potentially detect presumed germline pathogenic variants (PGPVs) in genes associated with hereditary diseases, and these are called germline findings. CGP contributes to the identification of germline variants in patients without a history of hereditary cancers. Approximately 25% of all ovarian cancers are caused by an inherited genetic factor. Detection of PGPVs leads to the identification of at-risk relatives and assessment and management of current and future cancers in patients and their relatives. Germline findings provide an opportunity to administer optimal molecular-targeted drugs. The presence of pathogenic variants *BRCA1/2* or homologous recombination repair deficiency confers sensitivity to poly (ADP-ribose) polymerase inhibitors in patients. Patients harboring DNA mismatch repair deficiency are highly sensitive to immune checkpoint inhibitors. Somatic genetic findings from CGP should be interpreted carefully, especially in patients with ovarian cancer. Improvement in the proportion of patients who undergo confirmatory germline genetic testing is an urgent task in the era of precision oncology.

Keywords Comprehensive cancer genomic profiling · Hereditary cancer · Germline findings · Presumed germline pathogenic variant(s) · *BRCA1/2* · Homologous recombination deficiency · Genetic counseling

---

M. Okazawa-Sakai (✉)
Department of Cancer Genomic Medicine, National Hospital Organization Shikoku Cancer Center, Matsuyama, Ehime, Japan

Department of Gynecologic Oncology, National Hospital Organization Shikoku Cancer Center, Matsuyama, Ehime, Japan
e-mail: sakai.mika.jt@mail.hosp.go.jp

D. Aoki et al. (eds.), *Practical Guide to Hereditary Breast and Ovarian Cancer*,
https://doi.org/10.1007/978-981-99-5231-1_11

1 Introduction

Comprehensive cancer genomic profiling (CGP) is being rapidly integrated into oncology practice with the evolution of next-generation sequencing (NGS) technology. While the major goal of CGP is to identify variants with potential therapeutic implications, CGP can potentially detect pathogenic germline variants in genes known to be associated with hereditary diseases, which are called germline findings [1]. Germline findings can have implications in treatment determination, risk assessment, and management of cancer in patients and their families. For example, patients harboring some germline pathogenic variants in homologous recombination repair genes, such as *BRCA1* and/or *BRCA2* (*BRCA1/2*) and *ATM*, are highly predictive of response to poly (ADP-ribose) polymerase (PARP) inhibitors [2–6]. In addition, identification of individuals carrying a pathogenic germline variant in cancer predisposition genes, such as *BRCA1/2*, allows for the prevention and early detection of future cancers [7]. Several guidelines have been established for clinicians to determine which patients should be referred to genetic specialists through CGP [1, 8–11].

Currently, genetic testing for germline and/or somatic *BRCA1/2* is essential for the care of ovarian cancer patients [7, 12, 13]. Previous studies have revealed a high frequency of heritable genetic conditions—approximately 18–24% of all ovarian cancers [14–17]. The increased detection of potentially clinically significant germline pathogenic variants has given rise to the need for an optimal approach to germline findings in care of ovarian cancer. This chapter describes the handling of germline findings recognized in CGP and their contribution to the identification of a family of hereditary ovarian cancers.

2 How to Evaluate Hereditary Predisposition to Cancer in Patients Undergoing CGP

2.1 *Presumed Germline Pathogenic Variants*

NGS technology permits the characterization of large amounts of DNA sequences and sequence variants detected in the tumor, including both somatic variants acquired during cancer development and germline variants. Although the origin of the variants is difficult to determine using tumor-only sequencing assays, germline pathogenic variants are inferred from CGP results without direct analysis of germline DNA. CGP-detected pathogenic variants of potential germline origin are called presumed germline pathogenic variants (PGPVs) [1].

Previous studies have revealed that 3–17% of patients undergoing cancer genomic profiling tests carry germline pathogenic variants [18–24]. Since June 2019, CGP tests have been reimbursed by the Japanese National Health Insurance System for cancer patients with unknown primary sites, rare tumors, or solid tumors refractory to standard treatment. Previous reports on CGP findings in core hospitals in Japan have demonstrated that 2–17% of patients have a tumor harboring PGPVs [25, 26]. In Japan, CGP testing has been performed in clinical practice using one of the following NGS-based panels: FoundationOne® CDx Cancer Genomic Profile (Foundation Medicine, MA, USA), OncoGuide™ NCC Oncopanel System (Sysmex, Kobe, Japan), or FoundationOne® Liquid CDx (Foundation Medicine, MA, USA) (as of May 5, 2022).

The FoundationOne CDx Cancer Genomic Profile is a tumor-only testing using DNA isolated from formalin-fixed paraffin-embedded (FFPE) tumor tissue specimens (tissue-based test). The PGPVs can be inferred from this test accordingly (Table 11.1). The OncoGuide NCC Oncopanel System is a tumor-normal paired test with germline variant subtraction. FFPE sections and peripheral blood are collected from the same patients, and the tumor and normal DNA are analyzed simultaneously. In this test, germline variants detected in normal DNA are subtracted from those detected in the tumor DNA. This panel also has a specific program to identify pathogenic germline variants using NGS data obtained from peripheral blood DNA. Pathogenic germline variants are detected in some genes associated with cancer predisposition (Table 11.1). The FoundationOne Liquid CDx Cancer Genomic Profile is a circulating tumor cell-free DNA (ctDNA)-based (liquid-based) test. The PGPVs can be inferred from this test accordingly (Table 11.1).

2.2 *Genes Recommended for Reporting to Patients*

Several guidelines have been published to assist clinicians in determining which patients should be referred to genetic specialists based on the CGP results. In 2012, the American College of Medical Genetics and Genomics (ACMG) published a minimum list of genes for which germline variants should be reported by clinical laboratories [27]. This list was updated in 2017 and 2021, respectively, and it finally

Table 11.1 Cancer genomic profiling tests and germline findings

Panel	Testing method	Germline findings
FoundationOne® CDx Cancer Genomic Profile	Tumor-only testing	PGPVs can be inferred
OncoGuide™ NCC Oncopanel System	Tumor-normal paired testing	
	• With germline variant subtraction	Any pathogenic germline variants may be invisible
	• With specific program of some genes associated with cancer predisposition	Pathogenic germline variants can be detected in the applicable genes
FoundationOne® Liquid CDx Cancer Genomic Profile	Circulating tumor DNA testing	PGPVs can be inferred

PGPVs presumed germline pathogenic variants

included 73 genes [28, 29] (Fig. 11.1; genes included in a green circle). The European Society of Medical Oncology (ESMO) and the French Society of Predictive and Personalized Medicine (SFMPP) have recommended lists of genes for inclusion in reports for germline findings [10, 11] (Fig. 11.1; genes included in purple and blue circles, respectively). In addition, the National Comprehensive Cancer Network (NCCN) guidelines specify genes for which the presence of germline pathogenic variants require specific management [7, 30] (Fig. 11.1; genes described in bold letters). The guidelines were modified to select genes recommended for reporting to patients and for being adopted in Japan [8] (Fig. 11.1; genes described in red letters).

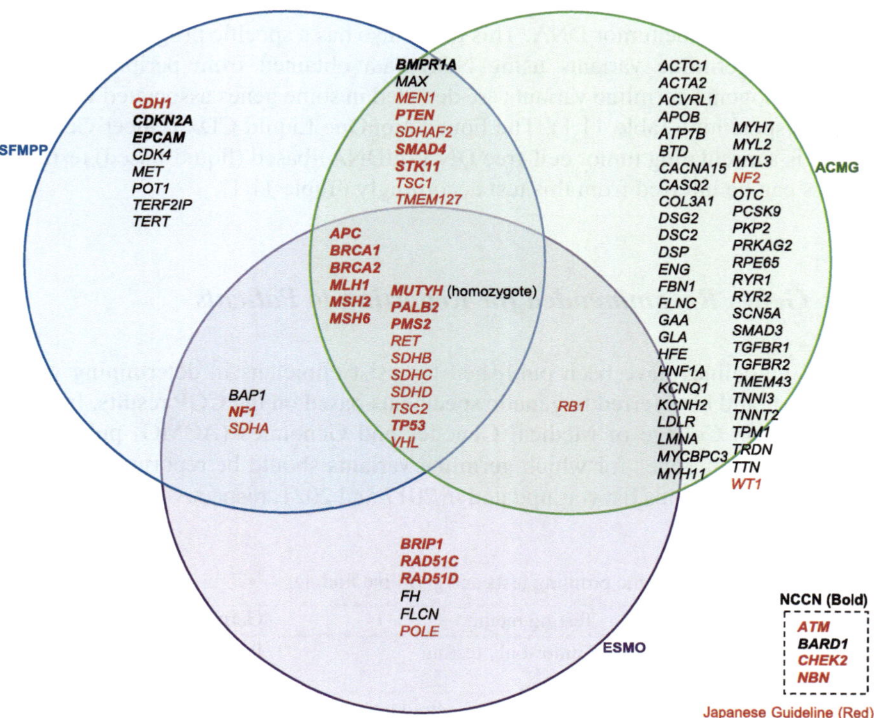

Fig. 11.1 Genes recommended for return of results. The lists of genes recommended for the return of results in ACMG (genes included in a green circle) [29], ESMO (genes included in a purple circle) [10], SFMPP (genes included in a blue circle) [11], NCCN guidelines (genes described in bold letters) [7, 30], and the Japanese Guideline (genes described in red letters) [8] are compared accordingly. The genes shown in the SFMPP part of this figure are "class 1 genes," which are defined as those for which information given to patients is recommended by the SFMPP [11]. This figure is created by the author. SFMPP, French Society of Predictive and Personalized Medicine; ACMG, American College of Medical Genetics and Genomics; ESMO, European Society of Medical Oncology; NCCN, National Comprehensive Cancer Network

2.3 Interpretation of Variants for Clinical Significance

The variants detected through CGP are evaluated based on the following public databases to annotate the pathogenicity: Catalog of Somatic Mutations in Cancer (COSMIC) [31], cBioPortal [32], and Clinical Interpretations of Variants in Cancer (CIViC) for somatic variants [33], and ClinVar for the relationship between germline variants and diseases [34]. The pathogenicity of the variant is classified using a five-tier system according to the guidelines established by ACMG and the Association for Molecular Pathology: benign, likely benign, variant of uncertain significance, likely pathogenic, and pathogenic [35]. Only pathogenic or likely pathogenic variants are reported accordingly.

2.4 Genetic Counseling–Referral Flow in CGP-Performed Patients

Figure 11.2 summarizes the genetic counseling-referral flow in CGP-performed patients as described in the guidelines [1, 8–10].

Generally, somatic pathogenic variants are not frequently detected in *BRCA1/2*, *PALB2*, *MSH2*, and *MSH6* by tumor DNA sequencing. Especially, almost 80% of pathogenic variants in *BRCA1/2* are of germline origin, i.e., high germline-conversion rates [10]. Therefore, all of pathogenic variants detected in *BRCA1/2* should be consider as PGPVs. In contrast, when pathogenic variants are detected in genes other than *BRCA1/2*, the variant allele frequency (VAF) supports clinicians to recognize the variants as PGPVs since the VAF of heterozygous germline variants generally ranges from 30% to 70% in tissue-based tests [10] and approximately 50% in liquid-based tests [36]. The ESMO presented the criteria for suspecting germline origin through tumor-only testing to achieve ≥10% germline-conversion rate per gene; targeted variants considered germline findings were restricted to those with VAF >30% (single nucleotide variants) or > 20% (small insertions/deletions) [10]. Indeed, liquid-based tests are more informative than tissue-based tests in the detection of PGPVs using VAF; however, the utility for the screening of germline variants remains unclear due to discrepancies in interpretation between somatic and germline sequence variants and technical limitations in tumor DNA sequencing to detect a broad spectrum of pathogenic variants that cause inherited disease predisposition [1].

Importantly, the VAF in tissue-based tests depends on the tumor purity, DNA ploidy, and local copy number [37]. If pathogenic or likely pathogenic variants are detected in genes that are commonly mutated in cancer, such as *APC*, *NF1*, *PTEN*, *RB1*, *STK11*, and *TP53*, reassessment of personal and family history is recommended (e.g., polyposis in APC or neurofibromas in NF1) [1].

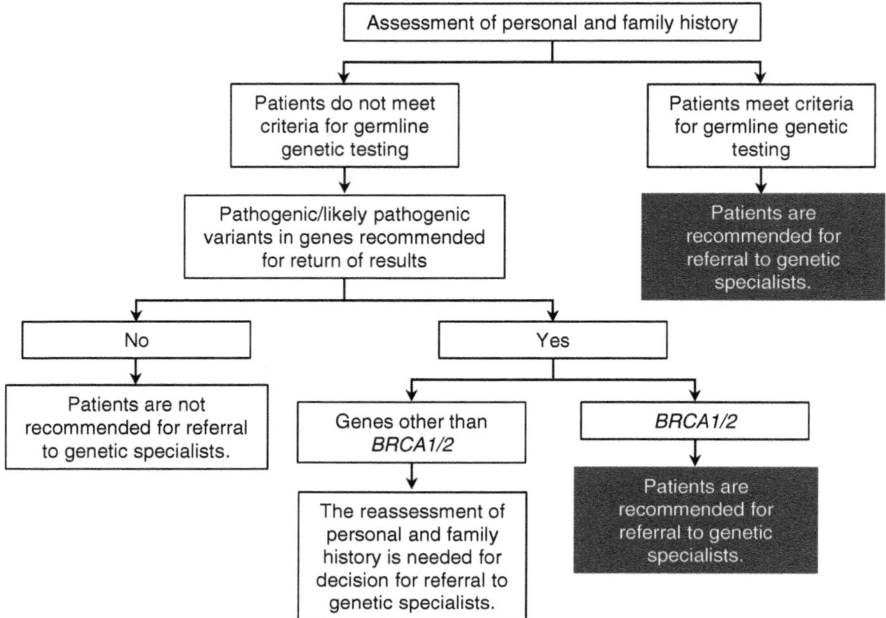

Fig. 11.2 Summary for referral to genetic specialists based on personal and family history and results of CGP. First, any patient for whom personal and/or family history meets the criteria for germline genetic testing is recommended for referral to genetic specialists. Second, when *BRCA1/2* pathogenic/likely pathogenic variants are detected, genetic specialists should be considered regardless of the tumor type and variant allele frequency (VAF). Lastly, when pathogenic/likely pathogenic variants in genes on the lists of guidelines for the return of results except for *BRCA1/2* are detected, a reassessment of personal and family history is needed for referral to genetic specialists, especially *APC*, *NF1*, *PTEN*, *RB1*, *STK11*, and *TP53*. Exclusion or confirmation of germline origin using VAF alone is not recommended in such patients. This figure is created by the author referring to the guidelines [1, 8–10]

2.5 *Personal and Family History in the Evaluation for Hereditary Predisposition*

The frequency of PGPVs fluctuates according to cancer type, tumor sample quality, tumor purity, somatic copy number alterations, genes analyzed in cancer genomic profiling, variant type, and testing method of cancer genomic profiling [1, 22, 37]. A recent large cohort study analyzed tumor and blood massive parallel sequencing data from 21,333 cancer patients and demonstrated that tumor-only sequencing failed to detect 10.5% of clinically actionable pathogenic germline variants in cancer susceptibility genes [22]. Therefore, germline genetic testing should be considered for patients with a personal and family history of hereditary cancer, but no PGPVs (Fig. 11.2). Exclusion or confirmation of germline origin using VAF alone is not recommended in such patients.

Personal and family history are essential for the identification of an individual with a risk of inherited predisposition to malignancy or other diseases. The American Society of Clinical Oncology (ASCO) recommends that the family history of patients with cancer should be assessed at the initial visit and reassessed periodically [38]. Generally, a hereditary cancer predisposition is considered if the patient displays an early age of cancer onset, multiple affected relatives with cancer on the same side of the family, or multiple primary tumors. Historically, a threshold for triggering germline test of 10%, which was based on personal and family history of cancer, has been widely adopted from the recommendation of the UK National Health Service [39]. Since approximately 25% of all ovarian cancers are caused by a heritable genetic condition [13], recent guidelines demonstrate that germline genetic testing for *BRCA1/2* should be conducted for all patients with ovarian cancer at initial diagnosis [7, 13, 40–43].

Finally, it should be noted that a normal or negative result for tumor sequencing is not equivalent to a normal/negative germline result [1]. Sequencing of germline DNA is the most sensitive approach, and sequencing of tissue DNA possibly misses almost 5% of germline pathogenic variants [5]. Germline genetic testing is still recommended for patients with ovarian cancer, even if tumor-only testing shows no *BRCA1/2* pathogenic variant. Multigene panel testing for germline sequencing that includes *BRCA1/2*, other homologous recombination repair genes, and DNA mismatch repair genes would be useful for such patients [13].

3 Clinical Utility of Germline Findings for Patients with Ovarian Cancer

3.1 Cancer Treatment

Screening is ineffective for ovarian cancer, and most patients are diagnosed with advanced disease. The standard therapy for patients with ovarian cancer has consists of cytoreductive surgery followed by platinum-based chemotherapy for almost 20 years [44]. While the response rate for first-line platinum-based chemotherapy is ~80%, most patients will experience recurrence within 2 years [44]. The introduction of PARP inhibitors has transformed treatment of ovarian cancer [44]. Inhibition of PARP generates single-stranded DNA breaks and accumulates double-stranded breaks, which require homologous recombination repair. Germline or somatic *BRCA1/2* mutation is strongly predictive for efficacy of PARP inhibitors in patients with ovarian cancer [5, 45]. Moreover, approximately 50% of tissue in ovarian cancer has HRD, and the efficacy of PARP inhibitors was confirmed in the patients harboring HRD or platinum-sensitivity in the tumor [46–51]. Currently, olaparib and niraparib are approved for maintenance therapy after response to platinum-based chemotherapy on the first-line treatment and the platinum-sensitive recurrent setting, i.e., the recurrence occurred at >6 months after the last platinum administration, in Japan (as of May 5, 2022) (Table 11.2). Germline *BRCA1/2* testing and HRD assays have become routine for ovarian cancer treatment (Table 11.2).

The frequency of microsatellite instability (MSI)-high or mismatch repair deficiency in ovarian cancer is not very high, ranging from 3–12% [52–55]. A meta-analysis revealed an overrepresentation of nonserous histology in ovarian cancer with mismatch repair deficiency [55]. These data support testing for MSI status or mismatch repair determined from tumor tissue in patients with ovarian cancer, especially for those with nonserous histology. Patients with MSI-high or MMR deficiency are responded to immune checkpoint blockade. Pembrolizumab, an anti-programmed death 1 (PD-1) inhibitor, has been approved for patients with MSI-high, unresectable, or metastatic cancer in Japan.

3.2 *Identification of a Family of Hereditary Ovarian Cancer*

Germline pathogenic variants of *BRCA1/2* are associated with an increased risk of breast, ovarian, pancreatic, and prostate cancers. The cumulative lifetime risk of ovarian cancer is 44% in *BRCA1* and 17% in *BRCA2* mutation carriers [56]. Risk-reducing bilateral salpingo-oophorectomy (RRSO) decreases the incidence of ovarian cancer and reduces 68% of overall mortality [57]; the role of RRSO in cancer prevention is well-established for women harboring germline pathogenic variants in

Table 11.2 Approvals for PARP inhibitors for patients with ovarian cancer in Japan (as of May 5, 2022)

PARP inhibitor	Biomarker	Setting		Pivotal trials supporting the approval
Olaparib	Germline or somatic *BRCA*m	First-line	Maintenance treatment of patients with advanced disease who have response to platinum-based CT	SOLO-1 [5]
	Tumor *BRCA*m or HRD	First-line	Combination with bevacizumab for first-line maintenance treatment of patients with advanced disease who have response to platinum-based CT	PAOLA-1 [48]
	None	Recurrent	Maintenance treatment of patients who have response to platinum-based CT	Study19 [49] SOLO-2 [45]
Niraparib	None	First-line	Maintenance treatment of patients with advanced disease who have response to platinum-based CT	PRIMA [51]
	None	Recurrent	Maintenance treatment of patients who are response to platinum-based CT	NOVA [50]
	Tumor *BRCA*m or HRD	Recurrent	Patients with disease progression longer than 6 months after response to the last platinum-based CT	QUADRA [47]

PARP poly (ADP-ribose) polymerase, *BRCA*m positive for *BRCA1/2* mutations, *HRD* homologous recombination repair deficiency, *CT* chemotherapy

BRCA1/2. The detection of pathogenic variants of *BRCA1/2* through CGP is critical not only for providing genome-matched therapy to the patient but also for identifying a family with *BRCA*-related hereditary breast and ovarian syndrome (HBOC). This can lead to the initiation of cascade testing and life-saving management for at-risk relatives.

Among patients with hereditary ovarian cancers, almost three-quarters are caused by germline alterations in *BRCA1/2*; the remaining quarter are caused by genes associated with homologous recombination repair or DNA mismatch repair [15]. The risk of ovarian cancer associated with mutations in moderate-risk genes, including *BRIP1*, *RAD51C*, *RAD51D*, *MLH1*, *MSH2*, and *MSH6*, is variable [58]. Germline pathogenic variants of DNA mismatch repair genes (*MLH1*, *MSH2*, and *MSH6*) are associated with Lynch syndrome, and the cumulative lifetime risk of ovarian cancer is estimated: 4–20% for *MLH1*, 8–38% for *MSH2*, and 1–13% for *MSH6* [58]. The estimated cumulative lifetime risk of ovarian cancer is 6–12% for carriers of a *BRIP1* pathogenic variant, 11% for carriers of a *RAD51C*, and 13% for carriers of a *RAD51D* [7]. Women harboring pathogenic germline variants in these genes are recommended to discuss RRSO [7].

Molecular stratification for treatment has been the standard of care for a broad range of cancer types, and the routine use of CGP will be a daily practice for ovarian cancer in future [59]. The identification of PGPVs is an entry point for genetic counseling for ovarian cancer patients, and it leads to the identification of a family with HBOC.

4 Barriers for Confirmatory Germline Genetic Testing

A recent Japanese study revealed that 14% of patients had PGPVs, but only 42% of these patients received genetic counseling [25]. One of the reasons for not undergoing germline genetic testing is patient death shortly after disclosure [25]. This indicates the need to ensure appropriate timing of CGP, shorten its turnaround time, and quickly refer patients to genetic specialists.

In Japan, genetic counseling, confirmatory germline genetic testing, surveillance, and risk-reducing surgery are not covered by the national health insurance system for all women, which is a barrier for them to undergo genetic counseling and confirmatory germline genetic testing. Interestingly, a Canadian study suggested that most patients who underwent CGP were interested in knowing their germline status [60]. Easier access to genetic medical services, including financial support, is required to improve the proportion of patients undergoing genetic counseling and germline genetic testing.

5 Conclusion

CGP contributes to the identification of germline variants in patients without a history of hereditary cancers. Germline findings identified through CGP can have implications for the assessment and management of future primary cancer risk, family risk assessment and guidance, and personalized treatment determination. Somatic genetic findings from CGP should be interpreted carefully, especially in patients with ovarian cancer. Improvement in the proportion of patients who undergo confirmatory germline genetic testing is an urgent task in the era of precision oncology.

References

1. Li MM, Chao E, Esplin ED, Miller DT, Nathanson KL, Plon SE, et al. Points to consider for reporting of germline variation in patients undergoing tumor testing: a statement of the American College of Medical Genetics and Genomics (ACMG). Genet Med. 2020;22(7):1142–8. https://doi.org/10.1038/s41436-020-0783-8.
2. de Bono J, Mateo J, Fizazi K, Saad F, Shore N, Sandhu S, et al. Olaparib for metastatic castration-resistant prostate cancer. N Engl J Med. 2020;382(22):2091–102. https://doi.org/10.1056/NEJMoa1911440.
3. Golan T, Hammel P, Reni M, Van Cutsem E, Macarulla T, Hall MJ, et al. Maintenance olaparib for germline BRCA-mutated metastatic pancreatic cancer. N Engl J Med. 2019;381(4):317–27. https://doi.org/10.1056/NEJMoa1903387.
4. Litton JK, Rugo HS, Ettl J, Hurvitz SA, Goncalves A, Lee KH, et al. Talazoparib in patients with advanced breast cancer and a germline BRCA mutation. N Engl J Med. 2018;379(8):753–63. https://doi.org/10.1056/NEJMoa1802905.
5. Moore K, Colombo N, Scambia G, Kim BG, Oaknin A, Friedlander M, et al. Maintenance olaparib in patients with newly diagnosed advanced ovarian cancer. N Engl J Med. 2018;379(26):2495–505. https://doi.org/10.1056/NEJMoa1810858.
6. Robson M, Im SA, Senkus E, Xu B, Domchek SM, Masuda N, et al. Olaparib for metastatic breast cancer in patients with a germline BRCA mutation. N Engl J Med. 2017;377(6):523–33. https://doi.org/10.1056/NEJMoa1706450.
7. NCCN Clinical Practice Guidelines in Oncology. Genetic/Familial High-Risk Assessment: Breast, Ovarian, and Pancreatic Version 1.2022. https://www.nccn.org/professionals/physician_gls. Accessed 16 May 2022.
8. Kyoto University Genetic Counseling Course (Japanese). http://sph.med.kyoto-u.ac.jp/gccrc/kouroukosugi.html. Accessed 16 May 2022.
9. DeLeonardis K, Hogan L, Cannistra SA, Rangachari D, Tung N. When should tumor genomic profiling prompt consideration of germline testing? J Oncol Pract. 2019;15(9):465–73. https://doi.org/10.1200/JOP.19.00201.
10. Mandelker D, Donoghue M, Talukdar S, Bandlamudi C, Srinivasan P, Vivek M, et al. Germline-focussed analysis of tumour-only sequencing: recommendations from the ESMO Precision Medicine Working Group. Ann Oncol. 2019;30(8):1221–31. https://doi.org/10.1093/annonc/mdz136.
11. Pujol P, Vande Perre P, Faivre L, Sanlaville D, Corsini C, Baertschi B, et al. Guidelines for reporting secondary findings of genome sequencing in cancer genes: the SFMPP recommendations. Eur J Hum Genet. 2018;26(12):1732–42. https://doi.org/10.1038/s41431-018-0224-1.

12. American College of Obstetricians and Gynecologists. Hereditary cancer syndromes and risk assessment. ACOG Committee Opinion No. 793. Obstet Gynecol. 2019;134:143–9.
13. Konstantinopoulos PA, Norquist B, Lacchetti C, Armstrong D, Grisham RN, Goodfellow PJ, et al. Germline and somatic tumor testing in epithelial ovarian cancer: ASCO guideline. J Clin Oncol. 2020;38(11):1222–45. https://doi.org/10.1200/JCO.19.02960.
14. Enomoto T, Aoki D, Hattori K, Jinushi M, Kigawa J, Takeshima N, et al. The first Japanese nationwide multicenter study of BRCA mutation testing in ovarian cancer: CHARacterizing the cross-sectionaL approach to Ovarian cancer geneTic TEsting of BRCA (CHARLOTTE). Int J Gynecol Cancer. 2019;29(6):1043–9. https://doi.org/10.1136/ijgc-2019-000384.
15. Hirasawa A, Imoto I, Naruto T, Akahane T, Yamagami W, Nomura H, et al. Prevalence of pathogenic germline variants detected by multigene sequencing in unselected Japanese patients with ovarian cancer. Oncotarget. 2017;8(68):112258–67. https://doi.org/10.18632/oncotarget.22733.
16. Norquist BM, Harrell MI, Brady MF, Walsh T, Lee MK, Gulsuner S, et al. Inherited mutations in women with ovarian carcinoma. JAMA Oncol. 2016;2(4):482–90. https://doi.org/10.1001/jamaoncol.2015.5495.
17. Walsh T, Casadei S, Lee MK, Pennil CC, Nord AS, Thornton AM, et al. Mutations in 12 genes for inherited ovarian, fallopian tube, and peritoneal carcinoma identified by massively parallel sequencing. Proc Natl Acad Sci U S A. 2011;108(44):18032–7. https://doi.org/10.1073/pnas.1115052108.
18. Mandelker D, Zhang L, Kemel Y, Stadler ZK, Joseph V, Zehir A, et al. Mutation detection in patients with advanced cancer by universal sequencing of cancer-related genes in tumor and normal DNA vs guideline-based germline testing. JAMA. 2017;318(9):825–35. https://doi.org/10.1001/jama.2017.11137.
19. Meric-Bernstam F, Brusco L, Daniels M, Wathoo C, Bailey AM, Strong L, et al. Incidental germline variants in 1000 advanced cancers on a prospective somatic genomic profiling protocol. Ann Oncol. 2016;27(5):795–800. https://doi.org/10.1093/annonc/mdw018.
20. Schrader KA, Cheng DT, Joseph V, Prasad M, Walsh M, Zehir A, et al. Germline variants in targeted tumor sequencing using matched normal DNA. JAMA Oncol. 2016;2(1):104–11. https://doi.org/10.1001/jamaoncol.2015.5208.
21. Seifert BA, O'Daniel JM, Amin K, Marchuk DS, Patel NM, Parker JS, et al. Germline analysis from tumor-germline sequencing dyads to identify clinically actionable secondary findings. Clin Cancer Res. 2016;22(16):4087–94. https://doi.org/10.1158/1078-0432.CCR-16-0015.
22. Terraf P, Pareja F, Brown DN, Ceyhan-Birsoy O, Misyura M, Rana S, et al. Comprehensive assessment of germline pathogenic variant detection in tumor-only sequencing. Ann Oncol. 2022;33(4):426–33. https://doi.org/10.1016/j.annonc.2022.01.006.
23. Dumbrava EI, Brusco L, Daniels M, Wathoo C, Shaw K, Lu K, et al. Expanded analysis of secondary germline findings from matched tumor/normal sequencing identifies additional clinically significant mutations. JCO Precis Oncol. 2019:3. https://doi.org/10.1200/PO.18.00143.
24. Jones S, Anagnostou V, Lytle K, Parpart-Li S, Nesselbush M, Riley DR, et al. Personalized genomic analyses for cancer mutation discovery and interpretation. Sci Transl Med. 2015;7(283):283ra53 https://doi.org/10.1126/scitranslmed.aaa7161.
25. Kikuchi J, Ohhara Y, Takada K, Tanabe H, Hatanaka K, Amano T, et al. Clinical significance of comprehensive genomic profiling tests covered by public insurance in patients with advanced solid cancers in Hokkaido, Japan. Jpn J Clin Oncol. 2021;51(5):753–61. https://doi.org/10.1093/jjco/hyaa277.
26. Sunami K, Ichikawa H, Kubo T, Kato M, Fujiwara Y, Shimomura A, et al. Feasibility and utility of a panel testing for 114 cancer-associated genes in a clinical setting: A hospital-based study. Cancer Sci. 2019;110(4):1480–90. https://doi.org/10.1111/cas.13969.
27. ACMG Board of Directors. Points to consider in the clinical application of genomic sequencing. Genet Med. 2012;14(8):759–61. https://doi.org/10.1038/gim.2012.74.
28. Kalia SS, Adelman K, Bale SJ, Chung WK, Eng C, Evans JP, et al. Recommendations for reporting of secondary findings in clinical exome and genome sequencing, 2016 update (ACMG SF

v2.0): a policy statement of the American College of Medical Genetics and Genomics. Genet Med. 2017;19(2):249–55. https://doi.org/10.1038/gim.2016.190.

29. Miller DT, Lee K, Chung WK, Gordon AS, Herman GE, Klein TE, et al. ACMG SF v3.0 list for reporting of secondary findings in clinical exome and genome sequencing: a policy statement of the American College of Medical Genetics and Genomics (ACMG). Genet Med. 2021; https://doi.org/10.1038/s41436-021-01172-3.
30. NCCN. Clinical Practice Guidelines in Oncology. Genetic/Familial High-Risk Assessment: Colorectal Version 1. 2021. https://www.nccn.org/professionals/physician_gls. Accessed 16 May 2022.
31. The Catalogue of Somatic Mutations In Cancer (COSMIC). https://cancer.sanger.ac.uk/cosmic Accessed 16 May 2022.
32. cBioPortal. https://www.cbioportal.org/. Accessed 16 May 2022.
33. Clinical Interpretations of Variants in Cancer (CIViC). https://civicdb.org/home. Accessed 16 May 2022.
34. ClinVar. https://www.ncbi.nlm.nih.gov/clinvar/. Accessed 16 May 2022.
35. Richards S, Aziz N, Bale S, Bick D, Das S, Gastier-Foster J, et al. Standards and guidelines for the interpretation of sequence variants: a joint consensus recommendation of the American College of Medical Genetics and Genomics and the Association for Molecular Pathology. Genet Med. 2015;17(5):405–24. https://doi.org/10.1038/gim.2015.30.
36. Slavin TP, Banks KC, Chudova D, Oxnard GR, Odegaard JI, Nagy RJ, et al. Identification of incidental germline mutations in patients with advanced solid tumors who underwent cell-free circulating tumor DNA sequencing. J Clin Oncol. 2018:JCO1800328. https://doi.org/10.1200/JCO.18.00328.
37. Sun JX, He Y, Sanford E, Montesion M, Frampton GM, Vignot S, et al. A computational approach to distinguish somatic vs. germline origin of genomic alterations from deep sequencing of cancer specimens without a matched normal. PLoS Comput Biol. 2018;14(2):e1005965. https://doi.org/10.1371/journal.pcbi.1005965.
38. Lu KH, Wood ME, Daniels M, Burke C, Ford J, Kauff ND, et al. American Society of Clinical Oncology Expert Statement: collection and use of a cancer family history for oncology providers. J Clin Oncol. 2014;32(8):833–40. https://doi.org/10.1200/JCO.2013.50.9257.
39. Familial Breast Cancer. Classification, care and managing breast cancer and related risks in people with a family history of breast cancer. London: National Institute for Health and Care Excellence: Guidelines; 2019.
40. Committee on Practice Bulletins-Gynecology, Committee on Genetics Society of Gynecologic Oncology. Practice bulletin No 182: hereditary breast and ovarian cancer syndrome. Obstet Gynecol. 2017;130(3):e110–e26. https://doi.org/10.1097/AOG.0000000000002296.
41. Randall LM, Pothuri B, Swisher EM, Diaz JP, Buchanan A, Witkop CT, et al. Multi-disciplinary summit on genetics services for women with gynecologic cancers: A Society of Gynecologic Oncology White Paper. Gynecol Oncol. 2017;146(2):217–24. https://doi.org/10.1016/j.ygyno.2017.06.002.
42. Colombo N, Sessa C, du Bois A, Ledermann J, McCluggage WG, McNeish I, et al. ESMO-ESGO consensus conference recommendations on ovarian cancer: pathology and molecular biology, early and advanced stages, borderline tumours and recurrent diseasedagger. Ann Oncol. 2019;30(5):672–705. https://doi.org/10.1093/annonc/mdz062.
43. Force USPST, Owens DK, Davidson KW, Krist AH, Barry MJ, Cabana M, et al. Risk assessment, genetic counseling, and genetic testing for BRCA-related cancer: US preventive services task force recommendation statement. JAMA. 2019;322(7):652–65. https://doi.org/10.1001/jama.2019.10987.
44. Mirza MR, Coleman RL, Gonzalez-Martin A, Moore KN, Colombo N, Ray-Coquard I, et al. The forefront of ovarian cancer therapy: update on PARP inhibitors. Ann Oncol. 2020;31(9):1148–59. https://doi.org/10.1016/j.annonc.2020.06.004.
45. Pujade-Lauraine E, Ledermann JA, Selle F, Gebski V, Penson RT, Oza AM, et al. Olaparib tablets as maintenance therapy in patients with platinum-sensitive, relapsed ovarian cancer

and a BRCA1/2 mutation (SOLO2/ENGOT-Ov21): a double-blind, randomised, placebo-controlled, phase 3 trial. Lancet Oncol. 2017;18(9):1274–84. https://doi.org/10.1016/S1470-2045(17)30469-2.
46. Cancer Genome Atlas Research N. Integrated genomic analyses of ovarian carcinoma. Nature. 2011;474(7353):609–15. https://doi.org/10.1038/nature10166.
47. Moore KN, Secord AA, Geller MA, Miller DS, Cloven N, Fleming GF, et al. Niraparib monotherapy for late-line treatment of ovarian cancer (QUADRA): a multicentre, open-label, single-arm, phase 2 trial. Lancet Oncol. 2019;20(5):636–48. https://doi.org/10.1016/S1470-2045(19)30029-4.
48. Ray-Coquard I, Pautier P, Pignata S, Perol D, Gonzalez-Martin A, Berger R, et al. Olaparib plus bevacizumab as first-line maintenance in ovarian cancer. N Engl J Med. 2019;381(25):2416–28. https://doi.org/10.1056/NEJMoa1911361.
49. Ledermann J, Harter P, Gourley C, Friedlander M, Vergote I, Rustin G, et al. Olaparib maintenance therapy in platinum-sensitive relapsed ovarian cancer. N Engl J Med. 2012;366(15):1382–92. https://doi.org/10.1056/NEJMoa1105535.
50. Mirza MR, Monk BJ, Herrstedt J, Oza AM, Mahner S, Redondo A, et al. Niraparib maintenance therapy in platinum-sensitive, recurrent ovarian cancer. N Engl J Med. 2016;375(22):2154–64. https://doi.org/10.1056/NEJMoa1611310.
51. Gonzalez-Martin A, Pothuri B, Vergote I, DePont Christensen R, Graybill W, Mirza MR, et al. Niraparib in patients with newly diagnosed advanced ovarian cancer. N Engl J Med. 2019;381(25):2391–402. https://doi.org/10.1056/NEJMoa1910962.
52. Cortes-Ciriano I, Lee S, Park WY, Kim TM, Park PJ. A molecular portrait of microsatellite instability across multiple cancers. Nat Commun. 2017;8:15180. https://doi.org/10.1038/ncomms15180.
53. Kelderman S, Schumacher TN, Kvistborg P. Mismatch repair-deficient cancers are targets for anti-PD-1 therapy. Cancer Cell. 2015;28(1):11–3. https://doi.org/10.1016/j.ccell.2015.06.012.
54. Murphy MA, Wentzensen N. Frequency of mismatch repair deficiency in ovarian cancer: a systematic review This article is a US Government work and, as such, is in the public domain of the United States of America. Int J Cancer. 2011;129(8):1914–22. https://doi.org/10.1002/ijc.25835.
55. Pal T, Permuth-Wey J, Kumar A, Sellers TA. Systematic review and meta-analysis of ovarian cancers: estimation of microsatellite-high frequency and characterization of mismatch repair deficient tumor histology. Clin Cancer Res. 2008;14(21):6847–54. https://doi.org/10.1158/1078-0432.CCR-08-1387.
56. Kuchenbaecker KB, Hopper JL, Barnes DR, Phillips KA, Mooij TM, Roos-Blom MJ, et al. Risks of breast, ovarian, and contralateral breast cancer for BRCA1 and BRCA2 mutation carriers. JAMA. 2017;317(23):2402–16. https://doi.org/10.1001/jama.2017.7112.
57. Eleje GU, Eke AC, Ezebialu IU, Ikechebelu JI, Ugwu EO, Okonkwo OO. Risk-reducing bilateral salpingo-oophorectomy in women with BRCA1 or BRCA2 mutations. Cochrane Database Syst Rev. 2018;8:CD012464. https://doi.org/10.1002/14651858.CD012464.pub2.
58. Liu YL, Breen K, Catchings A, Ranganathan M, Latham A, Goldfrank DJ, et al. Risk-reducing bilateral salpingo-oophorectomy for ovarian cancer: a review and clinical guide for hereditary predisposition genes. JCO Oncol Pract. 2022;18(3):201–9. https://doi.org/10.1200/OP.21.00382.
59. Mosele F, Remon J, Mateo J, Westphalen CB, Barlesi F, Lolkema MP, et al. Recommendations for the use of next-generation sequencing (NGS) for patients with metastatic cancers: a report from the ESMO Precision Medicine Working Group. Ann Oncol. 2020;31(11):1491–505. https://doi.org/10.1016/j.annonc.2020.07.014.
60. Stjepanovic N, Stockley TL, Bedard PL, McCuaig JM, Aronson M, Holter S, et al. Additional germline findings from a tumor profiling program. BMC Med Genet. 2018;11(1):65. https://doi.org/10.1186/s12920-018-0383-5.

Part VI
JOHBOC Registration

Registration Data of Japanese Organization of Hereditary Breast and Ovarian Cancer Till 2020

Mayuko Inuzuka, Masami Arai, and Seigo Nakamura

Abstract Clarifying the clinical and genetic characteristics of hereditary breast and ovarian cancer (HBOC) in Japan enables patients and their families to seek appropriate screening, diagnosis, and treatment options, thereby resulting in a lower breast cancer– and ovarian cancer–related mortality. We are compiling a nationwide registry by creating a database of families with HBOC syndrome in Japan at the Japanese Organization of Hereditary Breast and Ovarian Cancer with the aim of clarifying the clinical and genetic characteristics as well as improving the medical practice environment for such cases. We herein report a summary of the nationwide registry's September 2020 results.

In the 2020 compilation, a total of 28,846 (subjects and their family members) individuals from 93 institutions in Japan were registered. Of these, 7043 subjects were the first to undergo the *BRCA1/2* genetic test in their families. The *BRCA1/2* genetic test results revealed that 19.4% had the *BRCA1/2* pathogenic variant, whereas 5.3% had the variant of uncertain significance. In Japan, in April 2020, the insurance coverage for the *BRCA1/2* genetic test for breast or ovarian cancer patients with suspected HBOC was initiated. Consequently, as the financial burden of the *BRCA1/2* genetic tests was reduced, the number of subjects increased, thereby leading to the accumulation of data that was closer to the real conditions of the Japanese

M. Inuzuka
Department of Genetic Counseling, Showa University, Graduate School of Health Sciences, Tokyo, Japan
e-mail: inuzuka-m@cnt.showa-u.ac.jp

M. Arai
Department of Clinical Genetics, Juntendo University, Graduate School of Medicine, Tokyo, Japan
e-mail: ms-arai@juntendo.ac.jp

S. Nakamura (✉)
Division of Breast Surgical Oncology, Department of Surgery, Showa University School of Medicine, Tokyo, Japan
e-mail: seigonak@med.showa-u.ac.jp

D. Aoki et al. (eds.), *Practical Guide to Hereditary Breast and Ovarian Cancer*,
https://doi.org/10.1007/978-981-99-5231-1_12

people. It is the responsibility of those involved in the treatment of HBOC to continue the compilation of the nationwide registry, understand the current situation in Japan, and work toward improvement in the medical practice environment for HBOC.

Keywords *BRCA1/2* · HBOC · Database · JOHBOC · Registry

1 Japanese Organization of Hereditary Breast and Ovarian Cancer (JOHBOC) Registry

Overall, 5%–10% of all cancers are reported to have a genetic cause. Moreover, two types of genes, i.e., *BRCA1* and *BRCA2*, are known to be closely related to the onset of breast and ovarian cancers. When there is a pathogenic mutation in these genes that cause cancer, the risk of breast and ovarian cancer becomes extremely high, and this condition is called hereditary breast and ovarian cancer (HBOC) syndrome. Globally, there are many reports on HBOC, although it is well recognized that cancer onset varies by region and population. Understanding Japanese carriers of the *BRCA1/2* pathogenic variant allows for the provision of appropriate screening, diagnosis, and treatment options to patients and their families, which may lead to decreased mortality due to breast and ovarian cancers. Thus, in Japan, patients with HBOC should be registered with the nationwide registry in order to clarify the clinical and genetic characteristics of HBOC, such as the *BRCA1/2* variant characteristics and cancer penetrance of pathogenic variant carriers. We aim to develop a database of HBOC syndrome in Japan in order to better understand the clinical and genetic characteristics of HBOC as well as improve the medical practice environment for such patients.

Through the past data collection via this registry, the status of HBOC in Japan is gradually being clarified, and medical practice methods for HBOC are being improved. In Japan, Sugano et al. [1] performed the *BRCA1/2* genetic tests on 135 patients with breast or ovarian cancer in 2008, and they confirmed the presence of a pathogenic variant in 26.7% of the patients. Subsequently, Nakamura et al., in a group study by the Japanese Breast Cancer Society during 2012–2013 with the joint support from seven institutions [2], enrolled 320 patients with a strong family history of breast cancer. Among those who took the *BRCA1/2* genetic tests, Nakamura et al. confirmed the presence of the pathogenic variant and variant of unknown significance (VUS) in 30.7% and 6.2% patients, respectively. To follow-up with this study, we established the Japanese HBOC Consortium as a research organization and started the nationwide registry to further develop previous studies. In October 2013, the Japanese HBOC Consortium registry committee started planning for a nationwide registration system, and in 2015, trial registration was conducted at four institutions, in which registry committee members were affiliated in order to verify the registry system [3]. Finally, 846 families, with 965 subjects and their blood relatives (total: 3955 individuals), were enrolled in this study registry [4]. We contracted the participation in the nationwide registry in February 2016, when the procedure improved from the trial registration. In August 2016, 1557 families, with 1718 subjects and their blood relatives (total: 7118 individuals), were registered with the first nationwide registry [4]. Subsequently, in January 2019, the JOHBOC took over the registry that was previously managed by the Japanese HBOC Consortium. JOHBOC is an organization established in August 2016 to maintain and expand the medical practice systems for patients with or suspected HBOC as well as their families. It also aimed to contribute toward the improvement of preventive medicine and medical treatment for the citizens [5].

In the present registry, a jointly prepared family registry template by the JOHBOC and the National Clinical Database (NCD) was used. The treatment history and genetic counseling data of the subjects and their blood relatives were used to enter their clinical information. The study by Arai et al. [4] presents further information on the family registry template and registration items. The present study was a multicenter study conducted at various medical institutions with the approval of the ethics committee of each institution. The present study included the patients who underwent a genetic test, including the *BRCA1/2* test, at any of the participating institution. Each participating medical institution in this study, with the consent of the subjects, entered the *BRCA1/2* genetic test results and clinical information of the subjects and their blood relatives in the database. The entered data were anonymized by all the medical institutions before sending them to the NCD. The registered data of all cases are updated annually. Moreover, the data of the registered cases are summarized every year on the last day of August.

2 Summary of the Nationwide Registry Data Collection Results

Here, we report a summary of the nationwide registry's September 2020 data. By the end of August 2020, 28,846 (subjects and their blood relatives) individuals from 93 institutions across Japan were registered. Of those, 7997 underwent the *BRCA1/2* genetic test.

2.1 *Results of the* BRCA1/2 *Genetic Tests*

Among the 7997 subjects who underwent the *BRCA1/2* genetic test, 7043 were the first in their families to undergo the *BRCA1/2* genetic test. Overall, 5709 (81.1%) women had breast cancer, 803 (11.4%) women had ovarian cancer, 227 (3.2%) women had both breast and ovarian cancers, 48 (0.7%) men had breast cancer, and 256 (3.6%) individuals had neither breast nor ovarian cancer.

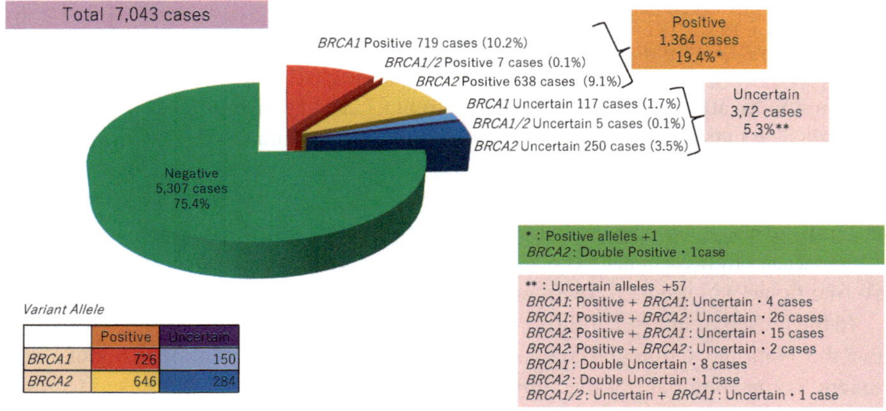

Fig. 1 The results of *BRCA1/2* genetic test of subjects who were the first to take the test in their family. Adapted from [4]

The *BRCA1/2* genetic test results of these 7043 subjects revealed that 1364 (19.4%) had the *BRCA1/2* pathogenic variant (Fig. 1). In total, 719 (10.2%), 638 (9.1%), and 7 (0.1%) patients had the *BRCA1*, *BRCA2*, and both *BRCA1* and *BRCA2* pathogenic variants, respectively. Among those with the *BRCA2* pathogenic variant, one patient had two *BRCA2* pathogenic variants. Regarding the VUS, 117 (1.7%), 250 (3.5%), and 5 (0.1%) patients had VUS of *BRCA1*, VUS of *BRCA2*, and VUS of both *BRCA1* and *BRCA2*, respectively. The detection rate of VUS of *BRCA1/2* was 5.3%.

2.2 *Clinical Background of the* BRCA1/2 *Pathogenic Variant Carriers*

Of the 7043 subjects who were the first to undergo the *BRCA1/2* genetic tests in the families, 1898 cases (26.9%) were women with only breast cancer with the initial onset at the age of ≥50 years, 3811 cases (54.1%) were women with only breast cancer with the initial onset at the age of <50 years, 803 cases (11.4%) were women with only ovarian cancer, 227 cases (3.2%) were women with both breast and ovarian cancers, 48 cases (0.7%) were men with breast cancer, regardless of the age at the initial onset, and 256 cases (3.6%) were individuals without breast or ovarian cancer. Table 1 depicts the medical history of 7043 subjects as well as their blood relatives by presenting the detection rates of the *BRCA1/2* pathogenic variants. Similar to the report by Arai et al. [4] that summarized the 2018 nationwide registry, the detection rate of the *BRCA1/2* pathogenic variants was higher in this study than that of the previously reported *BRCA1/2* mutations in non-Ashkenazi individuals in the US [6].

Moreover, the detection rate of *BRCA1/2* pathogenic variant in triple-negative breast cancer (TNBC) was reported in the present study. Among 1603 subjects

Table 1 Medical history and *BRCA1/2* pathogenic variant detection rate of the subject and blood relatives. Adapted from [4]

Family history	**Among 1st- or 2nd-degree relatives**				**No family history of breast or ovarian cancer**	**Family history in only 3rd-degree relatives**
Breast cancer < 50 years	−	+	−	+		
Ovarian cancer (at any age)	−	−	+	+		
Proband history Breast cancer ≥ 50 years	31/435 7.1%	54/346 15.6%	32/146 21.9%	13/35 37.1%	51/871 5.9%	4/65 6.2%
Breast cancer < 50 years	156/826 18.9%	251/790 31.8%	115/276 41.7%	67/118 56.8%	199/1645 12.1%	17/156 10.9%
Ovarian cancer at any age, no breast cancer	23/87 26.4%	23/46 50.0%	78/112 69.6%	13/15 86.7%	83/526 15.8%	2/17 11.8%
Breast cancer and ovarian cancer at any age	25/49 51.0%	13/26 50.0%	22/25 88.0%	9/10 90.0%	35/106 33.0%	3/11 27.3%
Male breast cancer at any age	1/4 25.0%	3/8 37.5%	0/1 0.0%	0/1 0.0%	2/30 6.7%	1/4 25.0%
No breast or ovarian cancer at any age	7/55 12.7%	11/86 12.8%	9/62 14.5%	7/26 26.9%	4/23 17.4%	0/4 0.0%

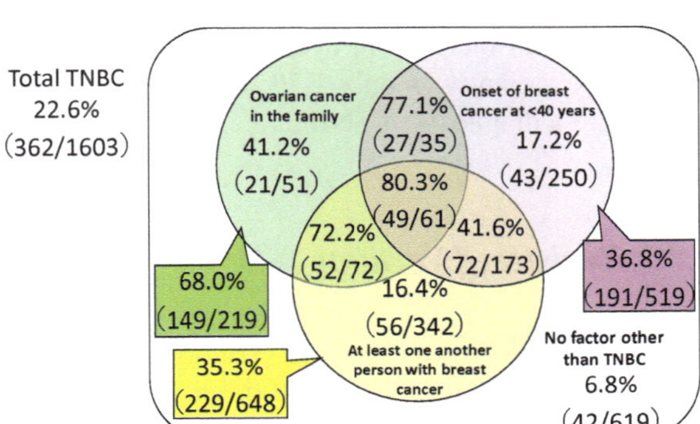

Fig. 2 *BRCA1* pathogenic variant detection rate in triple-negative breast cancer (TNBC)

diagnosed with TNBC, pathogenic mutation of either or both of the *BRCA* genes was detected in 449 (28.0%) patients; of these, 360 (80.2%) had the *BRCA1* pathogenic variant, 87 (19.4%) had *BRCA2* pathogenic variant, and 2 (0.4%) had both *BRCA1* and *BRCA2* pathogenic variants.

Here, the detection rate of the *BRCA1/2* pathogenic variant, considering factors (onset of breast cancer at <40 years, ovarian cancer in the family, and at least one another person with breast cancer) other than TNBC, is independently presented for *BRCA1* and *BRCA2*. Two cases with both the *BRCA1* and *BRCA2* pathogenic variants are included. Among all the TNBC cases, the detection rate for the *BRCA1* pathogenic variant among those without factors other than TNBC was 6.7% (Fig. 2). Furthermore, the detection rate was 17.8% in subjects with TNBC who developed breast cancer before the age of 40 years, 16.3% in those who had at least one other

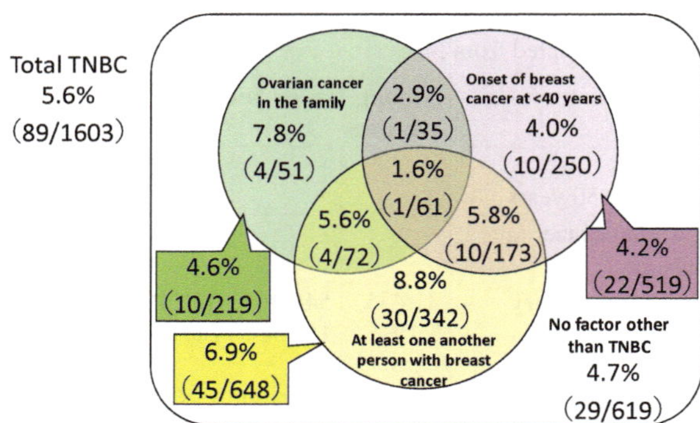

Fig. 3 *BRCA2* pathogenic variant detection rate in triple-negative breast cancer (TNBC)

family member with breast cancer, and 41.5% in those who had someone in their family with ovarian cancer. The detection rate of pathogenic variants increased as more factors overlapped. The detection rate of pathogenic variants was 83.1% when all three factors were present in addition to TNBC. Regarding the detection rate of *BRCA2* in all cases of TNBC, the rate of *BRCA2* in those without any factors other than TNBC was 4.6% (Fig. 3). The detection rate was 4.1% in subjects who developed breast cancer before the age of 40 years, 8.7% in those who had at least one other family member with breast cancer, and 7.5% in those who had someone in the family with ovarian cancer. Unlike *BRCA1*, the detection rate did not significantly change with an increase in overlapping factors. The detection rate of the *BRCA2* pathogenic variant was only 1.7% even when all the three factors were present in addition to a diagnosis of TNBC.

2.3 Registered Variants

In total, 208 *BRCA1* pathogenic variants were registered. Of these, 86 pathogenic variants were registered multiple times (Table 2). The most frequently registered variant was c.188 T > A (p.Leu63*) (177 cases), followed by c.2800C > T (p.Gln934*) (67 cases), c.2389_2390del (p.Glu797Thrfs*3) (25 cases), and c.5558A > G (p.tyr1853Cys) (18 cases). However, 148 *BRCA1* pathogenic variants were registered only once. In addition, 178 *BRCA2* pathogenic variants were registered. Of these 178 *BRCA2* pathogenic variants, 60 were registered multiple times (Table 3), with the most frequently registered variant being c.6952C > T (p.Arg2318*) (70 cases), followed by c.5576_5578del (p.lle1859Lysfs*3) (62 cases), c.9076C > T (p.Gln3026*) (44 cases), and c.8504C > A (p.Ser2835*) (31 cases). However, 118 *BRCA2* pathogenic variants were registered only once. Regarding VUS, 93 registrations were for *BRCA1*. Twenty VUS were registered multiple times. The most frequently registered variant was c.626C > T (p.Pro209Leu) (13 cases). Seventy-three VUS of *BRCA1* were registered only once. In total, 161 VUS of *BRCA2* were

Table 2 *BRCA1* pathogenic variant registered in more than 10 cases in the JOHBOC registry

Exon	HGVS cDNA	HGVS protein	Number of registered cases
5	c.188T > A	p.Leu63*	177
11	c.2800C > T	p.Gln934*	67
11	c.2389_2390del	p.Glu797Thrfs*3	25
24	c.5558A > G	p.Tyr1853C ys	18
3	c.131_132del	p.C ys44*	11

HGVS human genome variation society

Table 3 *BRCA2* pathogenic variant registered in more than 10 cases in the JOHBOC registry

Exon	HGVS cDNA	HGVS protein	Number of registered cases
13	c.6952C > T	p.Arg2318*	70
11	c.5576_5579del	p.Ile1859Lysfs*3	62
23	c.9076C > T	p.Gln3026*	44
20	c.8504C > A	p.Ser2835*	31
11	c.5645C > A	p.Ser1882*	25
10	c.1813del	p.Ile605Tyrfs*9	24
23	c.9117G > A	p.Pro3039=	19
10	c.1278del	p.Asp427Thrfs*3	15
18	c.8023A > G	p.Ile2675Val	15
11	c.2808_2811del	p.Ala938Profs*21	13
15	c.7558C > T	p.Arg2520*	12
20	c.8589dup	p.Ala2864Serfs*5	11

HGVS human genome variation society

registered. Of these, 41 VUS were registered multiple times, and the most frequently registered variant was c.53G > A (p.Arg18His) (28 cases). However, 120 VUS of *BRCA2* were registered only once.

3 Recent JOHBOC Data Results

We present some recently published research topics based on the JOHBOC nationwide registry data.

Sekine et al. [7] calculated the prevalence of ovarian and breast cancers as well as the ratio of ovarian cancer to breast cancer (number of patients with ovarian cancer: number of patients with breast cancer) with respect to common pathogenic variants in Japanese people, such as L63X and Q934X [8, 9] variants of *BRCA1*, which are known to be the founder pathogenic variants in Japan. They also examined whether there was any difference in the risk of cancer. With Q934X, the ratio of ovarian cancer to breast cancer was significantly higher than the overall *BRCA1*. With STOP799, the ratio was lower than the overall *BRCA1*. Both Q934X and

STOP799 were located in the ovarian cancer cluster region; moreover, there was a difference in the risk of ovarian cancer. Families with *BRCA1/2* pathogenic variants would benefit considerably if personalized counseling that takes variant type into account becomes available; hence, the results of the present study are useful.

Sekine et al. [10] analyzed the age of onset of ovarian cancer in Japanese women with the *BRCA1/2* pathogenic variant. Among those who underwent the *BRCA1/2* genetic test, those with the *BRCA1* pathogenic variant had a significantly younger age of onset of ovarian cancer. Moreover, those with the *BRCA2* pathogenic variant had a significantly older age of onset of ovarian cancer than those without the *BRCA1/2* pathogenic variant. Those with *BRCA2* pathogenic variant did not have an ovarian cancer onset before the age of 40 years. To the best of our knowledge, in Japan, these are the first pertinent scientific data to discuss the timing of risk-reducing salpingo-oophorectomy (RRSO).

Moreover, several such studies are ongoing.

4 Future Outlook

A summary of the 2020 nationwide registry, including 28,846 (subjects and their family members) individuals, was presented in this paper. In the 2019 summary, 15,612 (subjects and blood relatives) were registered from 62 institutions across Japan [4], implying that the 2020 summary included approximately twice the number of patients compared with the 2019 summary. From April 2020, the *BRCA1/2* genetic test for breast or ovarian cancer patients with suspected HBOC became covered by insurance in Japan. Since then, the number of patients undergoing the test has increased because of the reduced financial burden. In addition to the insurance coverage for the *BRCA1/2* genetic tests, coverage for RRM, RRSO, and magnetic resonance imaging assessments for patients with HBOC with a history of breast or ovarian cancer became effective. As some medical practice for HBOC is covered by insurance, we believe that the database accurately reflects the situation in Japan. With the current registry, from 2022, we will no longer register those without a pathogenic mutation or VUS in the *BRCA1/2* genetic test. After receiving the subject's consent to participate in the registry study, we plan to register the families of people who tested positive or had VUS with the *BRCA1/2* genetic test. The current database will be updated at least once a year. Moreover, the follow-ups with the registered subjects will be conducted, and the reports of blood relatives in relation to any cancer onset will be assessed. As the registry structure has a system that enables for long-term collection of follow-up data for registrants, we aim to provide a more accurate and valuable information with continuous stable functioning.

Additionally, insurance coverage for the *BRCA1/2* genetic test as a supplementary diagnostic test to determine the indication for a PARP inhibitor (olaparib) in HER2-negative inoperable or recurrent breast cancer with a history of chemotherapy began in 2018, and it became effective in 2019 for those with primary ovarian cancer. In January 2021, insurance coverage for the *BRCA1/2* genetic test was expanded as a supplementary diagnosis to determine the indication of olaparib for unresectable pancreatic and metastatic castration-resistant prostate cancers. In 2019, insurance coverage for cancer genomic profiling screening was approved for solid cancers without standard treatment or solid cancers where standard treatment was discontinued due to local progression or metastasis (including patients for whom the standard treatment is expected to be discontinued), thereby resulting in cases diagnosed with HBOC. Those of us working in HBOC medical practice have a responsibility to understand the importance of supplementary diagnostics and cancer genomic profiling in order to improve future medical practice for HBOC.

Meanwhile, medical practice systems for HBOC in Japan have faced many challenges, such as those related with the measurements for carriers of the *BRCA1/2* pathogenic variant without the onset of cancer. At present, regarding the the *BRCA1/2* genetic test for individuals without the onset of cancer, the possibility of risk-reducing resection and adequate surveillance of HBOC-related cancers are not covered by the insurance in Japan, thereby making it difficult to fully understand the situation due to the high financial burden. Hence, there is an urgent need to create an environment that allows individuals without an onset to be tested.

Acknowledgments JOHBOC Registration Committee [Committee]: Takayuki Enomoto (Niigata University Graduate School of Medical and Dental Sciences), Tadashi Nomizu (Hoshi General Hospital), Akihiro Sakurai (Sapporo Medical University), Masayuki Sekine (Niigata University Graduate School of Medical and Dental Sciences), Hiroyuki Nomura (Fujita Health University), Megumi Okawa (St. Luke's International Hospital), Hiraku Kumamaru (University of Tokyo), Takeo Kosaka (Keio University), and Masato Ozaka (Cancer Institute Hospital), [Secretariat of JOHBOC]: Eriko Shimamoto (Showa University) and Shiro Yokoyama (Showa University). Participating 93 institutions that entered their data with our 2020 registry: Showa University Hospital, St. Luke's International Hospital, the Cancer Institute Hospital of JFCR, Hoshi General Hospital Foundation, National Hospital Organization Shikoku Cancer Center, Hakuaikai Sagara Hospital, Hokkaido Cancer Center, Sapporo Medical University, Niigata University Medical and Dental Hospital, Keio University Hospital, Tazuke Kofukai Medical Research Institute Kitano Hospital, Kochi Medical School Hospital, Tawara IVF Clinic, Kochi Health Sciences Center, Nagoya University Hospital, St. Marianna University School of Medicine, Asahikawa Medical University, Aichi Cancer Center Hospital, Saku General Hospital Saku Medical Center, University of Tsukuba Hospital, Yokohama City University Hospital, Fujita Health University Hospital, Ehime University Hospital, Gunma Prefectural Cancer Center, Nagasaki University Hospital, Shinshu University Hospital, Juntendo University Hospital, Kitasato University Hospital, Kansai Medical University Hospital, Okazaki City Hospital, Okayama University Hospital, Saitama Medical University International Medical Center, Tokyo Medical Center, Tokyo Metropolitan Cancer and Infectious Diseases Center Komagome Hospital, Japanese Red Cross Ishinomaki Hospital, Nagoya City University Hospital, Kawasaki Medical School Hospital, Yamanashi Prefectural Central Hospital, Nara Medical University, Kanagawa Cancer Center, Shizuoka Cancer Center, Hiroshima Prefectural Hospital, Japanese Red Cross Aichi Medical Center Nagoya Daiichi Hospital, Tottori University Hospital, Kokura Medical Center, Nippon Medical School Musashikosugi Hospital, Ibaraki Prefectural Central Hospital, University of Fukui Hospital, Hokkaido University Hospital, Osaka University Hospital, Japanese Red Cross Yamaguchi Hospital, Kyoto University Hospital, Japanese Red Cross Medical Center, Shonan Memorial Hospital, Shizuoka General Hospital, Hiroshima University Hospital, National Hospital Organization Kyoto Medical Center, Shiroyama Hospital, National Hospital Organization Osaka National Hospital, Kitakyushu Municipal Medical Center, Iwate Medical University, the Jikei University Hospital, Tokuyama Central Hospital, Yamagata Prefectural Central Hospital, Nihonkai

General Hospital, Fukushima Medical University, Japanese Red Cross Nagoya Daini Hospital, Yamaguchi University Hospital, Tosei General Hospital, Seirei Hamamatsu General Hospital, Saitama Red Cross Hospital, Faculty of Medicine University of Miyazaki Hospital, JA Hiroshima General Hospital, Saiseikai Hyogo Prefectural Hospital, JCHO Saitama Medical Center, Osaka Prefectural Hospital Organization Osaka International Cancer Institute, Ishikawa Prefectural Central Hospital, Hyogo Medical University, Aizawa Hospital, Shizuoka City Shizuoka Hospital, JCHO Sagamino Hospital, Tokushima University Hospital, Kanazawa University Hospital, Ogaki Municipal Hospital, Oikawa Hospital, Nippon Medical School Hospital, Toyota Memorial Hospital, National Center for Global Health and Medicine, Hyogo Prefectural Amagasaki General Medical Center, Kagawa Prefectural Central Hospital, Hamamatsu University Hospital, Japanese Red Cross Wakayama Medical Center, Fujisawa City Hospital.

References

1. Sugano K, Nakamura S, Ando J, Takayama S, Kamata H, et al. Cross-sectional analysis of germline BRCA1 and BRCA2 mutations in Japanese patients suspected to have hereditary breast/ovarian cancer. Camcer Sci. 2008;99(10):1967–76.
2. Nakamura S, Takahashi M, Tozaki M, Nakayama T, Nomizu T, et al. Prevalence and differentiation of hereditary breast and ovarian cancers in Japan. Breast Cancer. 2015;22(5):462–8.
3. Arai M, Yokoyama S, Watanabe C, Yoshida R, Kita M, et al. Genetic and clinical characteristics in Japanese hereditary breast and ovarian cancer: first report after establishment of HBOC registration system in Japan. J Hum Gent. 2018;63(4):447–57.
4. Arai M, The Registration Committee of the Japanese Organization of Hereditary Breast and Ovarian Cancer (JOHBOC). Hereditary breast and ovarian cancer. Hereditary breast and ovarian cancer (HBOC) database in Japan. In: Nakamura S, Aoki D, Miki Y, editors. Hereditary breast and ovarian cancer molecular mechanism and clinical practice. Tokyo: Springer; 2021. p. 243–57.
5. Japan Hereditary Breast Cancer Ovarian Cancer Comprehensive Medical Care System Organization. https://johboc.jp/. Accessed 31 May 2022.
6. Frank TS, Deffenbaugh AM, Reid JE, Hulick M, Ward BE, et al. Clinical characteristics of individuals with germline mutations in BRCA1 and BRCA2: analysis of 10,000 individuals. J Clin Oncol. 2002;20:1480–90.
7. Sekine M, Enomoto T, Arai M, Yokoyama S, Nomura H, et al. Correlation between the risk of ovarian cancer and BRCA recurrent pathogenic variants in Japan. J Hum Genet. 2022;67(5):267–72.
8. Sekine M, Nagata H, Tsuji S, Hirai Y, Fujimoto S, et al. Mutational analysis of BRCA1 and BRCA2 and clinicopathologic analysis of ovarian cancer in 82 ovarian cancer families: two common founder mutations of BRCA1 in Japanese population. Clin Cancer Res. 2001;7:3144–50.
9. Sekine M, Nagata H, Tsuji S, Hirai Y, Fujimoto S, et al. Localization of a novel susceptibility gene for familial ovarian cancer to chromosome 3p22-p25. Hum Mol Genet. 2001;10:1421–9.
10. Sekine M, Enomoto T, Arai M, Den H, Nomura H, et al. Differences in age at diagnosis of ovarian cancer for each BRCA mutation type in Japan: optimal timing to carry out risk-reducing salpingo-oophorectomy. J Gynecol Oncol. 2022; https://doi.org/10.3802/jgo.2022.33.e46.

GPSR Compliance

The European Union's (EU) General Product Safety Regulation (GPSR) is a set of rules that requires consumer products to be safe and our obligations to ensure this.

If you have any concerns about our products, you can contact us on ProductSafety@springernature.com

In case Publisher is established outside the EU, the EU authorized representative is:

Springer Nature Customer Service Center GmbH
Europaplatz 3
69115 Heidelberg, Germany

Batch number: 10370708

Printed by Printforce, the Netherlands